Studies in Applied Philosophy, Epistemology and Rational Ethics

Volume 66

Studies in Applied Philosophy, Epistemology and Rational Ethics (SAPERE) publishes new developments and advances in all the fields of philosophy, epistemology, and ethics, bringing them together with a cluster of scientific disciplines and technological outcomes: ranging from computer science to life sciences, from economics, law, and education to engineering, logic, and mathematics, from medicine to physics, human sciences, and politics. The series aims at covering all the challenging philosophical and ethical themes of contemporary society, making them appropriately applicable to contemporary theoretical and practical problems, impasses, controversies, and conflicts.

Book proposals can be sent to the Series Editor, Prof. Lorenzo Magnani at lmagnani@unipv.it. Please include: A short synopsis of the book or the introduction chapter A Table of Contents The CV of the lead author/editor(s) For more information, please contact the Editor-in-Chief at lmagnani@unipv.it

Our scientific and technological era has offered "new" topics to all areas of philosophy and ethics – for instance concerning scientific rationality, creativity, human and artificial intelligence, social and folk epistemology, ordinary reasoning, cognitive niches and cultural evolution, ecological crisis, ecologically situated rationality, consciousness, freedom and responsibility, human identity and uniqueness, cooperation, altruism, intersubjectivity and empathy, spirituality, violence. The impact of such topics has been mainly undermined by contemporary cultural settings, whereas they should increase the demand of interdisciplinary applied knowledge and fresh and original understanding. In turn, traditional philosophical and ethical themes have been profoundly affected and transformed as well: they should be further examined as embedded and applied within their scientific and technological environments so to update their received and often old-fashioned disciplinary treatment and appeal. Applying philosophy individuates therefore a new research commitment for the 21st century, focused on the main problems of recent methodological, logical, epistemological, and cognitive aspects of modeling activities employed both in intellectual and scientific discovery, and in technological innovation, including the computational tools intertwined with such practices, to understand them in a wide and integrated perspective. **Studies in Applied Philosophy, Epistemology and Rational Ethics** means to demonstrate the contemporary practical relevance of this novel philosophical approach and thus to provide a home for monographs, lecture notes, selected contributions from specialized conferences and workshops as well as selected Ph.D. theses. The series welcomes contributions from philosophers as well as from scientists, engineers, and intellectuals interested in showing how applying philosophy can increase knowledge about our current world. Indexed by SCOPUS, zbMATH, SCImago, DBLP. All books published in the series are submitted for consideration in Web of Science.

Simona Chiodo

Technology and the Overturning of Human Autonomy

Springer

Simona Chiodo
Department of Architecture and Urban
Studies (DAStU)
Politecnico di Milano
Milano, Italy

ISSN 2192-6255 ISSN 2192-6263 (electronic)
Studies in Applied Philosophy, Epistemology and Rational Ethics
ISBN 978-3-031-26158-9 ISBN 978-3-031-26159-6 (eBook)
https://doi.org/10.1007/978-3-031-26159-6

This Springer imprint is published by the registered company Springer Nature Switzerland AG
The registered company address is: Gewerbestrasse 11, 6330 Cham, Switzerland

[W]ho can view things in their connection is
a dialectician;
he who cannot is not

Plato

Acknowledgments

I am grateful to both colleagues and friends, from philosophers to scientists to a variety of scholars in between, with whom I have reflected upon the issues of my work, and to anonymous reviewers, who have helped me refine my reflection, especially in the following publications as the basis of my work, which is their further development, and whose reviewers I also thank: the monograph *Technology and Anarchy. A Reading of Our Era* (Lexington Books-The Rowman & Littlefield Publishing Group, 2020), the journal articles *From je ne sais quoi to Quantified Self. A Philosophical Agenda* ("Studi di Estetica", 2021), *Adding Wisdom to Computation. The Task of Philosophy Today* ("Metaphilosophy", 2022), *Human Autonomy, Technological Automation (and Reverse)* ("AI & Society. Journal of Knowledge, Culture and Communication", 2022) and *The Quantified Self as Epistemological Anarchism* ("Philosophia. Philosophical Quarterly of Israel", 2022) and the book chapter *Trading Human Autonomy for Technological Automation* (in S. Lindgren, *Handbook of Critical Studies of Artificial Intelligence*, Edward Elgar Publishers, forthcoming in 2023).

I also thank Lorenzo Magnani for accepting this book for publication in the book series "Studies in Applied Philosophy, Epistemology and Rational Ethics (SAPERE)".

Introduction

The aim of my work is to try, at least, to address the task Plato gives philosophers through his words I choose as epigraph: "who can view things in their connection is a dialectician; he who cannot is not" (Plat. *Resp.* 537 c). Contemporary philosophers frequently focus on issues as circumscribed as scientists' issues are circumscribed—which may be promising, I believe, only if they do not stop exercising the capacity from which philosophy itself arises: the capacity to "connect" the dots through a kind of synthesis, i.e. vision of the whole, after having analysed them. And the key reasons why "connecti[ng]" the dots through a kind of synthesis is essential are, first, that the dots can make more sense when they are understood as parts of the whole and, second, that it is only from a kind of synthesis, i.e. vision of the whole, that novel visions of the whole can possibly arise (for instance, I can have a novel vision of my whole life only if I move from focusing on one of its dimensions to focusing on one of its dimensions' relationship with my whole life's whatever dimension, synchronically and diachronically).

The two key reasons described above underpin my work as my attempt to understand current technologies by "connecti[ng]" them with a greater scenario of relationships: first, relationships between current technologies and historical cornerstones that can make more sense of them and, second, and especially, relationships between current technologies and philosophy's classic questions, which can go even further in making more sense of them. More precisely, as the title of my work says, the key question I shall address as one of philosophy's most classic questions is the following: what is the relationship between current technologies and human autonomy? And the core of my answer shall be the following: their relationship is a matter of overturning—the relationship between current technologies and human autonomy is a matter of overturning in the sense that it is a matter of trading human autonomy for technological automation (for complex and even quite touching reasons, as I shall argue).

Trading humans' autonomy for technology's automation does not mean at all that technology can be autonomous in the same way as humans can be autonomous. Actually, technology as we know it cannot be autonomous in the ways in which the history of Western philosophy thinks of autonomy, in that, from rational self-determination (especially in ancient philosophy) to the capacity to give oneself laws (especially in modern philosophy) to individualism (especially in contemporary philosophy), technology lacks what is most essential to be autonomous, i.e. its own reasons—technology as we know it can have no reasons but humans' reasons. Yet, technology can be autonomous in the sense that it can be thought of, defined and used by us as autonomous, as I shall show through several cases in point—more precisely, I shall argue that technology can be autonomous as our perfect way to escape from our own autonomy by making technology be our perfect scapegoat, i.e. our own autonomy's perfect avatar. And the consequence of our making technology be our own autonomy's perfect avatar is that, even though technology cannot be actually autonomous, i.e. autonomous in the same way as humans can be autonomous, we can actually lose our own autonomy—we can actually become more and more automated, i.e. technology's automation's perfect avatars.

I shall develop the core of my answer through two parts. In the first part of my work, I shall especially, but not exclusively, work on analysing two cases in point to focus on technology's philosophical meanings. Before analysing them, in Chap. 1, I shall start reflecting upon the relationship between technology and the overturning of human autonomy by getting insights from the figure of Prometheus from past to present as the most powerful symbol developed by Western culture to express the meaning of technology to humans, who are represented as constitutively technological. Actually, in Western culture, technology defines humans as capable of surviving and even ruling over nature (of which they are part). Moreover, surprisingly enough, what even the cradle of Western culture teaches us is that speaking of technology means speaking of the relationship between humans and the divine—technology is constitutively human and, at the same time, constitutively making humans try to increasingly obtain divine prerogatives. Thus, technology plays a key role in defining crucial hierarchies, from power (between rulers and ruled) to ontology (between creators and created), and its great power can even make it a player in the challenge between humans and the divine (which is what happens moving from ancient culture to modern and contemporary culture). Finally, the cradle of Western culture gives us an essential insight to understand humans' relationship with technology: humans seem to constitutively risk using technology without considering what is plural and common, specifically without giving it priority over what is singular and individual. Moving from the ancient Prometheus (as he is represented especially by Hesiod, Aeschylus, Plato and Ovid) to the modern Prometheus (as he is represented especially by Mary Shelley), the relationship between humans and technology becomes more challenging: technology makes humans even equal the divine, in that they are represented both as creators of living human matter from nonliving matter and as even challenging death—technology is represented as what makes humans obtain divine prerogatives. And technology itself increasingly becomes more powerful, specifically something that itself challenges both humans and the divine. Finally, moving from the modern Prometheus to a possible contemporary Prometheus, the relationship between humans and technology becomes even more challenging: technology makes humans even surpass the divine, in that humans create a kind of technology that itself is divine by obtaining omnipresence, omniscience, omnipotence and inscrutability, as especially current information technologies show—more precisely, technology is represented as what makes humans create a kind of divine that is perfectly immanent: on the one hand, it is an artefact, i.e. something perfectly immanent, but, on the other hand, it itself is omnipresent, omniscient, omnipotent and inscrutable, which are the most typical prerogatives of the divine as something transcendent. For instance, whenever we google something, we contribute to create, through our data, a kind of divine that is perfectly immanent by being, on the one hand, an artefact, i.e. something perfectly immanent, and, on the other hand, omnipresent (by being everywhere: it is always with us, from our desks to our pockets), omniscient (by knowing everything: it knows not only the answers to our questions but also ourselves by tracking us), omnipotent (by having more and more power: it has more and more power over us, from shaping our worldviews to shaping our decisions and, consequently, our actions) and inscrutable (algorithms as the basis of a great number of current technologies can be defined as black boxes, in that the reasons why a given input results in a given output can be inscrutable).

The first outcome of my initial philosophical reading shall be that, even from an etymological perspective, speaking of an immanent divine means speaking of a radically anarchic move, as I shall extensively argue in the other chapters. More precisely, speaking of anarchism means speaking of radical absence of a ruler, i.e. radical rulerlessness. The human creation of a technological divine may be thought of as a radically anarchic move for at least two key reasons: first, humans succeed in creating a technological divine that is totally immanent and, second, the totally immanent technological divine does not succeed in being an actual ruler in itself, in that actually ruling seems to require actual expertise, starting with expertise in understanding relationships between contexts and data, and actual expertise seems to require a hierarchical difference from a totally immanent technological randomness. Thus, several current technologies mark our era as a radical overturning of hierarchical differences, starting with the fundamental dissolution of the divine not only into the human but also into the technological, which is the key reason why our technological era may be read as the most radical form of anarchism we have ever experienced in Western culture.

Moreover, the second outcome of my initial philosophical reading shall be that, as I shall extensively argue in the other chapters, overturning hierarchical differences, starting with the immanentisation of the ruler as what our technological era increasingly shows, means overturning human autonomy, which literally means self-given rules, as one of the most essential cornerstones of the definition itself of human identity and dignity in Western culture. Actually, if ruling itself experiences a crisis, then what is the destiny of human autonomy in our technological era? More precisely, if our technological era is characterised by an increasing loss of rules, which are increasingly replaced by a kind of immanentised automated randomness correlating with a kind of technological bureaucracy's inscrutability (which may be read as the most radical form of anarchism we have ever experienced in Western culture), then what is the destiny of human autonomy precisely as what requires rules, i.e. self-given rules, from its etymology to philosophy?

In Chaps. 2 and 3, I shall work on the analysis of two cases in point, chosen as great representatives of current technologies, to answer the questions. In Chap. 2, I shall focus on the philosophical reading of autonomous technologies, especially autonomous vehicles. Surprisingly enough, if we pay attention to the current way we use language, then we realise that we increasingly shift the word "autonomy" from defining ourselves to defining technologies. Several authors work on autonomy as something possessed by technology, but I shall work on something quite different: autonomy as something possessed by humans and, consequently, the philosophical meaning of the shift from humans' autonomy to technology's autonomy, on the one hand, and the philosophical meaning of the shift from technology's automation to humans' automation, on the other hand. I shall read the shifts described above as the attempt to shift the dramatic burden of responsibility from humans to technology by making the former automated and the latter autonomous. But the actually advantageous shifts also have potential disadvantages, which are crucial: the increasing technological immanentisation of rulers and rules makes us risk falling into the randomness that results from nothing but empirical sums, which technology's ease of computing surely strengthen, as several autonomous technologies show. More precisely, we should be aware that technology's ease of computing, together with the divine prerogatives (omnipresence, omniscience, omnipotence and inscrutability) it increasingly obtains, equally increases a severe risk: taking mere statistics, i.e. kinds of empirical sums, instead of rules. Yet, again, rules resulting from merely particular dimensions can give us nothing but the randomness of contingency, and even of idiosyncrasy. Moreover, statistics taken instead of rules can do nothing but work as something heteronomously imposed on our autonomy and freedom. Thus, why do we trade our autonomy for technology's automation? My working hypothesis is that autonomy started becoming its own worst enemy after its irresistible rise especially from modern Western culture to contemporary Western culture: what burden could be heavier than the burden of an exceedingly celebrated autonomy? More precisely, if autonomy even defines human identity and dignity, then what burden could be heavier than autonomy itself for humans? Thus, one of the key reasons why we increasingly trade our autonomy for technology's automation is precisely the following: the more our autonomy is on the rise, the heavier, and even unbearable, the burden we have to bear both epistemologically and ethically is—paradoxically enough, the more our autonomy is on the rise, the more we trade it for what can make us move from being causes of effects to being effects of causes. Is technology, specifically technology's automation, our perfect solution? The answer seems affirmative: if something goes wrong, then technology can be our perfect scapegoat to which we can shift our epistemological and ethical burdens, from the most dramatic ones to the easiest ones, which means that technology as humans' perfect scapegoat especially means that fault can more easily be technology's, no one's and, in the worst case scenario, common, but it cannot more easily be individual's—and shifting from ourselves to technologies nothing less than our most successful definition as humans means that a kind of paradoxical externalisation of our own core identity emerges.

In Chap. 3, I shall focus on the philosophical reading of self-tracking technologies and the quantified self. My working hypothesis is that they cannot be read as a continuation of individual autonomy's rise, but as a symptom of its crisis, in that they sabotage what underpins individual autonomy itself by stressing both a kind of radical particularisation that may be read, again, as anarchism and a kind of reshaping of humans' core identity that may be read, again, as a trade of autonomy for automation. More precisely, autonomy's irresistible rise seems to be followed by autonomy's equally irresistible crisis, which means that what we may actually want to obtain from technology is precisely to trade our (heavier) core identity for a (lighter) digital identity—again, our (heavier) autonomy for technology's (lighter) automation. As self-tracking technologies and the quantified self show, technology as what can shape our technological exoselves is a scapegoat through which, from when we should find the courage to act to when we should face our actions' irreducibly uncertain outcomes, we unburden ourselves of our heaviest burdens, from our sense of responsibility, in the best case scenario, to our sense of guilt, in the worst case scenario. Day after day, sometimes quite explicitly and sometimes quite implicitly, we somehow want our technological exoselves not to be the continuation of ourselves, but the solution of continuity with ourselves.

In the second part of my work, I shall especially work on "connecti[ng]" the dots through a kind of synthesis, i.e. vision of the whole, to focus on challenges of the future, specifically epistemological, human identity's and ethical. More precisely, the aim of the second part of my work shall be asking philosophical questions, which is one of the classic tasks of philosophy, even more than answering. Yet, through the plausible scenarios I shall describe, I shall try, at least, to start answering. In Chap. 4, I shall address epistemological challenges. The historical trajectory briefly described above, from ancient culture to contemporary culture, results in countless current technologies based on computation, which literally means speaking of *logos* as the form of rationality that, in Western culture, wins over whatever way to obtain reliable knowledge, from forms of irrationality to other forms of rationality. But current computing technologies, especially technological exoselves, lead us to the following question: if we increasingly externalise *logos* from ourselves to computing technologies, in that they process more data more quickly (and, especially, in that they bear our epistemological and ethical burdens), then what is the destiny of both our *logos* in particular and our rationality in general? The question implies at least two further issues I shall address: first, the issue of risking atrophying our rational capacities, which strictly correlate with our autonomy, and, second, the issue of facing uncertainty, which *logos* alone, both internalised and externalised, cannot solve, as I shall argue. My working hypothesis is that wisdom can serve as the most promising complementary form of rationality of *logos*, especially to address current computing technologies' epistemological challenges. Starting with the history of philosophy, which defines wisdom as self-awareness and awareness of lacks and obstacles, resilience and, especially, practice-oriented knowledge mastering domains that *logos* cannot master, we may develop wisdom's definition as the capacity to imagine a series of things, i.e. possible lacks and obstacles, possible ways both to be adaptable and to make a virtue of necessity and, especially, possible identifications of the whole, from values to far more extended contexts, especially in terms of long-term consequences. If our developed definition of wisdom makes sense, then the most essential reason why wisdom should be added to *logos* as its most promising complementary form of rationality is that it can make us act even though we have no certainty at all, in that it can give us reasons to act that can be valid even though they cannot be certain—and, in our era's complexity and uncertainty, having valid reasons to act even though we do not have certain reasons to act can even save us, from our health to our morality and from ourselves as single individuals to our society, as I shall argue. But is technology capable of wisdom? My answer is that, at least currently, technology does not seem capable of being wise, if we define wisdom as the capacity to imagine especially what we cannot know as at least probabilistically certain, i.e. what is uncertain. And the key reason why technology does not seem capable of being wise is that is seems to lack imagination, from both an epistemological perspective and an ethical perspective, as I shall argue.

In Chap. 5, I shall address human identity's challenges, starting with the following question: what may be the destiny of human identity if we increasingly move autonomy, as one of the most essential cornerstones of our definition itself as humans, from ourselves to technology? I shall continue to investigate the sense in which we can actually become more and more automated, i.e. technology's automation's perfect avatars, by exponentially trading our doubts, as ways to shape our own reasons that make sense of our lives, for technological efficiency, as a way to do more things more quickly. Current technologies are frequently clear symptoms of the shift from humans to technology of what has been defined for millennia as something essentially human not only in terms of prerogatives and capacities but also, and especially, in terms of what pleasantly makes sense of human life. Yet, until making sense of human life will constitutively be humans' key issue, there will be something unshiftable from humans to technology. But we risk undermining our making sense of our lives if we end up being more and more driven by technology's kinds of reasons, i.e. no one's reasons, whenever our actions result from something that stops coinciding with reasons and starts coinciding with causes, specifically technology's random causes, which happens whenever technology's outputs start being, meaningfully, autonomous, specifically autonomous from our inputs. I shall use something from the past as what may help us face the risk described above: the notion of autarky as self-sufficiency, which is precisely the core of the wise's life in several ancient philosophies, and, especially, which is the condition of possibility for human autonomy as the condition of possibility for human making sense of human life. And I shall especially argue that the key reason why the wisest thing we can do is not to shift whatever issue to solve from ourselves to technology is that, otherwise, we would risk losing our most crucial capacity—more precisely, we would risk losing our most crucial meta-capacity, specifically our meta-capacity to face uncertainty (which, day after day, more strongly characterises our era).

In Chap. 6, I shall address ethical challenges, starting with the following questions: if our era of globalisation, characterised by unprecedented complexity and uncertainty, may be the worst time to lose not only our autonomy in general but also our autonomy as underpinning our intersubjectivity in particular (precisely as what finally underpins our capacity to face global, i.e. shared, challenges), then what may be the philosophical meaning of the paradoxical overturning of our autonomy? And, especially, what may be, from an ethical perspective, the philosophical meaning of our paradoxical trade of our autonomy for technology's automation? I shall reflect upon cases in point, especially the relationship between technology and health and the relationship between technology and politics, in which technology seems to stress subjectivity again and again, which means ending up stressing a contingent, and even idiosyncratic, randomness into which whatever ruler finally dissolves. If it is true that, from an epistemological perspective, the notion of tailoring, specifically to single individuals, increasingly characterises our use of technology, it is also true that, from an ethical, and even political, perspective, several current technologies may be read as supported by the idea, and as equally supporting the idea, according to which we have more and more rights, and even hyperindividualistic rights, but less and less duties. More precisely, several autonomous technologies seem to primarily show that we have the right to do whatever we want to do, even though we lack expertise, and especially if we lack expertise, starting with the following series of technologies upon which I shall reflect as kinds of anarchic immanentisations of rulers: even though we cannot drive (from not having driving licence to other reasons), we can somehow drive (through autonomous vehicles), even though we cannot diagnose (not being doctors), we can somehow self-diagnose (through self-tracking technologies and the quantified self), even though we cannot work on the drafting of laws (not being jurists), we can somehow work on the drafting of laws (through a digital platform), even though we cannot design houses (not being architects), we can somehow design houses (through softwares), even though we cannot write articles (not being writers), we can somehow write articles (through softwares), even though we cannot be professionals at doing almost whatever, from cooking to coaching (being amateurs), we can somehow be professionals at doing almost whatever, from cooking to coaching (through social media platforms), even though we cannot be professionals at getting whatever kind of information (not being journalists), we can somehow be professionals at getting whatever kind of information (through Google). And so forth. Through the cases in point described above, I shall further argue that what we actually want, sometimes consciously and sometimes unconsciously, is to unburden ourselves of our autonomy as what has increasingly become our most unbearable burden in the last three centuries—and technology may be our best solution, in that automation itself is our perfect scapegoat, i.e. our own autonomy's perfect avatar, by unburdening us of our responsibilities both as single individuals and as a society.

And, even though we may seem victims of a process of which we increasingly lose control, we are far from being victims. My working hypothesis is that, conversely, we are active, sometimes consciously and sometimes unconsciously, and not passive—more precisely, we are active in dramatically, and even touchingly, trying to get by, by purposely creating technologies as perfect scapegoats, i.e. our own autonomy's perfect avatars, through which, from when we should find the courage to act to when we should face our actions' irreducibly uncertain outcomes, we unburden ourselves of our heaviest burdens, from our sense of responsibility, in the best case scenario, to our sense of guilt, in the worst case scenario.

Contents

About the Author

Simona Chiodo is a full professor of philosophy at Politecnico di Milano. She is doing research in epistemology, specifically on the relationship between reality and ideality, in philosophy of technology, focusing on understanding the philosophical meaning of current technologies, and in aesthetics, specifically on the notion of beauty. She is the author of *Technology and Anarchy. A Reading of Our Era* (Lexington Books-The Rowman & Littlefield Publishing Group, 2020).

Part I
Technology's Philosophical Meanings

Chapter 1
Contemporary Prometheus

1.1 Insights from the Past

If it is true that we cannot understand the present without considering the past, it is also true that whoever wants to understand the present relationship between humans and technology, at least in Western culture, should consider the historical development of the figure of Prometheus as the most powerful symbol developed by Western culture to express the meaning of technology to humans.[1] I myself extensively worked on the figure of Prometheus as the symbol of the Western relationship between humans and technology,[2] and, since I realised that it is actually an exceedingly powerful symbol to understand not only what happened in the past but also what happens in the present, I believe that the most promising way to start reflecting upon technology and the overturning of human autonomy is precisely to get insights from the figure of Prometheus from past to present. Thus, on the following pages, I shall briefly focus on key issues aiming at better understanding the present through the past by considering the ancient Prometheus as he is represented especially by Hesiod, Aeschylus, Plato and Ovid and the modern Prometheus as he is represented especially by Mary Shelley. Finally, I shall use the historical insights as clues to illuminate the meaning of the current relationship between ourselves and technology.

[1] References are several, from art to philosophy (see at least Kahn, 1970; Winner, 1977 and Dougherty, 2006. As far as the move from ancient culture to modern culture is concerned, see at least Finn, Guston & Robert, 2017 and Cambra-Badii, Guardiola & Banos, 2020).

[2] Especially in Chiodo (2020a).

S. Chiodo, *Technology and the Overturning of Human Autonomy*,
Studies in Applied Philosophy, Epistemology and Rational Ethics 66,
https://doi.org/10.1007/978-3-031-26159-6_1

Even the etymology of the word "Prometheus" is illuminating. Literally, the word "Prometheus" means "predicting", in that it derives from the Greek words πρó ("before") and μανθάνω ("knowing").[3] Prometheus' capacity to predict is not only what makes him "artful [...], full of various wiles" (Hes. *Theog.* 510), "wily-minded" (Hes. *Theog.* 520) and "very knowing" (Hes. *Theog.* 615) but also "helpful" (Hes. *Theog.* 610), specifically to humans. As Plato writes (see Plat. *Prot.* 320 d ff.), Prometheus fixes Epimetheus' mistake[4] by giving humans what would have defined them forever, i.e. Hephaestus' fire and Athena's wisdom in the arts ("Prometheus, in his perplexity as to what preservation he could devise for man, stole from Hephaestus and Athena wisdom in the arts together with fire—since by no means without fire could it be acquired or helpfully used by any—and he handed it there and then as a gift to man", Plat. *Prot.* 321 c–d)—which symbolically means that what would have defined humans forever is precisely technology. And the reason why Hephaestus' fire and Athena's wisdom in the arts, i.e. technology, are "helpful" to humans is that the former make the latter survive and even rule over nature (of which they are part). Plato writes that humans move from being "naked, unshod, unbedded, unarmed" (Plat. *Prot.* 321 c) to having the "skill to articulate speech and words, and to invent dwellings, clothes, sandals, beds, and the foods that are of the earth" (Plat. *Prot.* 322 a). Aeschylus is even more expressive when he makes Prometheus describe humans' condition before he helped them: "listen to the miseries that beset mankind—how they were witless before and I made them have sense and endowed them with reason. I will not speak to upbraid mankind but to set forth the friendly purpose that inspired my blessing. First of all, though they had eyes to see, they saw to no avail; they had ears, but they did not understand; but, just as shapes in dreams, throughout their length of days, without purpose they wrought all things in confusion. They had neither knowledge of houses built of bricks and turned to face the sun nor yet of work in wood; but dwelt beneath the ground like swarming ants, in sunless caves. They had no sign either of winter or of flowery spring or of fruitful summer, on which they could depend but managed everything without judgment, until I taught them to discern the risings of the stars and their settings, which are difficult to distinguish. Yes, and numbers, too, chiefest of sciences, I invented for them, and the combining of letters, creative mother of the Muses' arts, with which to hold all things in memory. I, too, first brought brute beasts beneath the yoke to be subject to the collar and the pack-saddle, so that they might bear in men's stead their heaviest burdens; and to the chariot I harnessed horses and made them obedient to the rein, to be an image of wealth and luxury. It was I and no one else who invented the mariner's flaxen-winged car that roams the sea. Wretched that I am—such are the arts I devised for mankind" (Aesch. *PB* 436–470). We may translate Aeschylus' expressions into philosophical words as follows: the meaning of Prometheus' help to humans is as crucial as the definition itself of their ontology, which moves from "swarming ants" to creatures

[3] Predicting is one of Prometheus' prerogatives, also stressed by his opposition to his brother Epimetheus, who is described as "of-erring-mind" by Hesiod (Hes. *Theog.* 510).

[4] Epimetheus had to give creatures the qualities that would have defined them forever, but he totally forgot humans.

that not only survive nature by having qualities that make them capable of thinking of technological solutions ("sense", "reason", "discern[ment]" and, consequently, "numbers" and "letters" in particular, together with "arts" in general) but also rule over nature by having qualities that make them capable of actually creating technological solutions ("the yoke", "the collar", "the pack-saddle", "the chariot", "the rein" and the "flaxen-winged car"). Thus, the philosophical meaning of Prometheus' help to humans is that technology, from its causes to its effects, is precisely what makes humans human (otherwise, humans would be "swarming ants", i.e. equal to other animals)—more precisely, humans are constitutively technological.[5]

Surprisingly enough, one of the most critical issues characterising the current debate on technology is perfectly solved by the cradle of Western culture. The answer to the question on humans' relationship with technology is that the latter is not the former's ontological addition (sometimes totally welcomed and sometimes totally rejected)—conversely, technology is humans' ontological essence: what even the cradle of Western culture teaches us is that technology is not only the reason why humans can rule over nature but also, and especially, the reason why humans can be human.

Aeschylus' words go far beyond references to the kind of technological solutions we have seen ("the yoke", "the collar", "the pack-saddle", "the chariot", "the rein" and the "flaxen-winged car"). In addition, most sophisticated technologies are symbolically represented, which even stress the idea according to which humans are constitutively technological. First, the capacity to predict, which is the crucial aim of both science and technology, emerges ("And I marked out many ways by which they might read the future, and among dreams I first discerned which are destined to come true; and voices baffling interpretation I explained to them, and signs from chance meetings. The flight of crook-taloned birds I distinguished clearly—which by nature are auspicious, which sinister—their various modes of life, their mutual feuds and loves, and their consortings; and the smoothness of their entrails, and what color the gall must have to please the gods, also the speckled symmetry of the liver-lobe; and the thigh-bones, wrapped in fat, and the long chine I burned and initiated mankind into an occult art. Also I cleared their vision to discern signs from flames, which were obscure before this", Aesch. *PB* 480–495). Second, medicine, which makes humans not only survive nature but also improve their quality of life, emerges ("if ever man fell ill, there was no defence—no healing food, no ointment, nor any drink—but for lack of medicine they wasted away, until I showed them how to mix soothing remedies with which they now ward off all their disorders", Aesch. *PB* 475–480). Third, raw materials, which make humans capable of creating artefacts, emerge ("Now as to the benefits to men that lay concealed beneath the earth—bronze, iron, silver, and gold—who would claim to have discovered them before me? No one, I know full well, unless he likes to babble idly. Hear the sum of the whole matter in the compass of one brief word—every art possessed by man comes from Prometheus", Aesch.

[5] It is no coincidence that Prometheus is a Titan, and the Titans are most ancient gods who even predate the Olympians ruled by Zeus (see their description as πρότεροι θεοί, transliterated as *proteroi theoi*, Hes. *Theog*. 424).

PB 500–505). The first issue, i.e. the capacity to predict, is exceedingly meaningful. As we have seen, the capacity to predict is inherent in Prometheus' nature from both an etymological perspective and an ontological perspective (predicting is one of Prometheus' prerogatives, see note 3)—which means that the capacity to predict is divine (being Prometheus a Titan). Yet, we have also seen that the capacity to predict is inherent in humans' nature, in that technology, which is humans' ontological essence, aims at predicting, from Aeschylus' expressive examples to the current algorithms we meaningfully define as predictive algorithms. Thus, the question we should ask ourselves is the following: why is the divine capacity to predict also characterising humans? More precisely, why are humans also characterised by something divine?

Plato makes us reflect upon the possible answer by giving us a crucial clue, when he specifies what follows, after having written that Prometheus gives humans Hephaestus' fire and Athena's wisdom in the arts: "And now that man was partaker of a divine portion, he, in the first place, by his nearness of kin to deity, was the only creature that worshipped gods, and set himself to establish altars and holy images; and secondly, he soon was enabled by his skill to articulate speech and words, and to invent dwellings, clothes, sandals, beds, and the foods that are of the earth" (Plat. *Prot*. 322 a). As we have seen, the last part of Plato's words tells us that humans are constitutively technological, and that technology is what makes humans survive and even rule over nature. As far as the first part of Plato's words is concerned, his crucial clue has to do with the relationship between what is human and what is divine. The first thing worth noting is that the former and the latter are not totally divided. Conversely, humans are characterised by something more than "nearness of kin to deity", in that they are "partaker[s] of a divine portion". Moreover, the second thing worth noting is that the reason why humans are "partaker[s] of a divine portion" is that Prometheus gives them Hephaestus' fire and Athena's wisdom in the arts, i.e. divine prerogatives—the reason why humans are "partaker[s] of a divine portion" is precisely technology.

Surprisingly enough, what the cradle of Western culture teaches us is not only that humans are constitutively technological but also, and especially, that speaking of technology means speaking of the relationship between humans and the divine (which is an issue that also characterises both the modern Prometheus and, especially, the contemporary Prometheus). More precisely, as we shall see, speaking of technology means speaking of the human way to challenge the divine in terms of trying to increasingly obtain divine prerogatives in addition to the capacities symbolised by Hephaestus' fire and Athena's wisdom in the arts—consequently, we may say that speaking of technology means speaking of humans trying to overturn their ontology by increasing its "divine portion" (which is an issue that seems to have quite paradoxical outcomes in our technological era, as we shall see).

The ancient Prometheus also shows us the phenomenon described above as follows. When Prometheus tries to further help humans by deceiving even Zeus,[6]

[6] "When the gods and mortal men were contending at Mekone, then did he [Prometheus] set before him [Zeus] a huge ox, having divided it with ready mind, studying to deceive the wisdom of Zeus.

who discovers his deception, Zeus meaningfully punishes humans, even though they are not guilty.[7] And both their being punished and their kind of punishment are meaningful: "from that time forward, ever mindful of the fraud, he [Zeus] did not give the strength of untiring fire to wretched mortal men, who dwell upon the earth" (Hes. *Theog.* 560). Two details are most instructive. First, humans are punished even though they are not guilty, which shows a kind of challenge between them (who would have had advantages from Prometheus) and Zeus (who would have had disadvantages from Prometheus)—again, humans are not totally divided from the divine, to which they even seem to pose a kind of threat. Second, the kind of threat they even seem to pose to the divine is neutralised by depriving them of fire—the most instructive detail is that, if humans challenge the divine by trying to obtain its prerogatives (which are symbolically represented as advantages from Prometheus, see note 6), then what can neutralise them is precisely their deprivation of technology.

Thus, Western culture arises from the following idea: technology is constitutively human and, at the same time, constitutively making humans try to increasingly obtain divine prerogatives. It is no coincidence that Ovid describes Prometheus' special way to qualify humans as follows: "Still missing was a creature finer than these [the animals], with a great mind, one who could rule the rest: man was born, whether fashioned from immortal seed by the Master Artisan who made this better world, or whether Earth, newly parted from Aether above, and still bearing some seeds of her cousin Sky, was mixed with rainwater by Titan Prometheus and moulded into the image of the omnipotent gods. And while other animals look on all fours at the ground he gave to humans an upturned face, and told them to lift their eyes to the stars" (Ovid *Met* I 77–87). Again, the special way in which humans are qualified, i.e. technology, makes them rule over nature ("rule the rest") and, especially, not only obtain something divine (being "moulded into the image of the omnipotent gods") but also try to increasingly obtain further divine prerogatives (being given "an upturned face" and, especially, being told "to lift their eyes to the stars", which symbolically represents an ontological dimension that is literally superior to humans' ontological dimension, as the divine is *par excellence*).

The challenge between humans and the divine, specifically the former's threat posed to the latter, seems to be hinted by Aeschylus' dialogue between Prometheus

For here, on the one hand, he deposited the flesh and entrails with rich fat on the hide, having covered it with the belly of the ox; and there, on the other hand, he laid down, having well-disposed them with subtle craft, the white bones of the ox, covering them with white fat. Then it was that the father of gods and men addressed him, 'Son of Iapetus, far-famed among all kings, how unfairly, good friend, you have divided the portions'. Thus spoke rebukingly Zeus, skilled in imperishable counsels. And him in his turn wily Prometheus addressed, laughing low, but he was not forgetful of subtle craft: 'Most glorious Zeus, greatest of ever-living gods, choose which of these your inclination within your breast bids you'. He spoke subtlety: but Zeus knowing imperishable counsels was aware, in fact, and not ignorant of his guile; and was boding in his heart evils to mortal men, which also were about to find accomplishment. Then with both hands he lifted up the white fat. But he was incensed in mind, and wrath came around him in spirit, when he saw the white bones of the ox arranged with guileful art" (Hes. *Theog.* 535–555).

[7] Prometheus is punished by Zeus after having taken fire back to humans, as Aeschylus masterfully tells us in his *Prometheus Bound*.

and Io: "'But now no limit to my tribulations has been appointed until Zeus is hurled from his sovereignty'. 'What! Shall Zeus one day be hurled from his dominion? [...] By whom shall he be despoiled of the sceptre of his sovereignty?'. 'By himself and his own empty-headed purposes'. 'In what way? Oh tell me, if there be no harm in telling'. 'He shall make a marriage that shall one day cause him distress'. 'With a divinity or with a mortal? If it may be told, speak out'. 'Why ask with whom? I may not speak of this'. 'Is it by his consort that he shall be dethroned?'. 'Yes, since she shall bear a son mightier than his father'. 'And has he no means to avert this doom?'. 'No, none—except me, if I were released from bondage'" (Aesch. *PB* 755–770). The identity of the "son mightier than his father" is not revealed, but it is worth noting, first, that there is a correlation between Zeus' descendent destiny, whose power decreases ("Zeus is hurled from his sovereignty"), and humans' ascendant destiny, whose power increases (see Aesch. *PB* 436–470, 475–505), and, second, that the balance of power is ruled by Prometheus ("'And has he no means to avert this doom?'. 'No, none—except me'")—the balance of power is ruled by the symbol of the relationship between humans and technology. Again, the idea from which Western culture arises has to do with the key role played by technology in defining crucial hierarchies, from power (between rulers and ruled) to ontology (between creators and created, as we shall see). But we should ask a further question, which is even more complex: what is the hierarchical role played by technology in the challenge between the human and the divine? More precisely, what is technology's power? Is it the power of something ruled? Or is it the power of something that can rule? And what is technology's ontology? Is it the ontology of something created? Or is it the ontology of something that can create?

The historical development of the figure of Prometheus shall help us answer the complex questions. Yet, the ancient Prometheus can also give us meaningful insights. Zeus punishes humans not only by depriving them of fire but also by giving them Pandora and the jar: "and he [Zeus] called this woman Pandora, because all the gods who abide in Olympus gave her as a gift a pain for grain-eating men. But when the gods completed this deception of sheer doom, against which there is no remedy, father Zeus sent the famed Argos-killer to Epimetheus, the swift messenger of the gods, bringing the gift. Nor did Epimetheus take notice how Prometheus had told him never to accept a gift from Zeus the Olympian, but to send it right back, lest an evil thing happen to mortals. But he accepted it, and only then did he take note in his *noos* that he had an evil thing on his hands" (Hes. *Op.* 80–89). Before humans were given Pandora and the jar, "the various kinds of humanity lived on earth without evils and without harsh labour, without wretched diseases that give disasters to men" (Hes. *Op.* 90–92). But the greatest tragedy happened to humans when Pandora "took the great lid off the jar and scattered what was inside. She devised baneful anxieties for humankind. The only thing that stayed within the unbreakable contours of the jar was *Elpis* [Hope]. It did not fly out" (Hes. *Op.* 94–97). What the myth symbolically shows us is most interesting to understand the hierarchical role played by technology in the challenge between the human and the divine. Zeus punishes humans by depriving them of something and by giving them something at the same time. On the one hand, he deprives humans of fire, i.e. technology. On the other hand, he gives humans "pain", "sheer doom", "evil", "harsh labour", "wretched diseases", "disasters" and "baneful anxieties": Zeus gives humans what they try to solve through technology—Zeus' punishment stresses that what is most powerful for humans is technology. Interestingly enough, what humans try to solve through technology is given them with something else: "The only thing that stayed within the unbreakable contours of the jar was *Elpis* [Hope]. It did not fly out". Thus, on the one hand, there is "sheer doom, against which there is no remedy". But, on the other hand, there is still "*Elpis*", i.e. hope—the myth seems to symbolically show us that the challenge between the human and the divine is not even stopped by something "against which there is no remedy". Thus, technology turns out to be exceedingly powerful for humans—and whatever is exceedingly powerful risks moving from something passive to something active, which means moving from playing a role in the challenge between two players, i.e. the human and the divine, to being a player.

Thus, a meaningful insight given us by the ancient Prometheus is that technology's greatest power can even make it a player in the challenge between not only the human and the divine but also the technological—technology itself can challenge both power hierarchies (by being not only something ruled but also something that can rule) and ontological hierarchies (by being not only something created but also something that can create).

Before moving from the ancient Prometheus to the modern Prometheus, we should reflect upon one last insight. Even though humans have something exceedingly powerful, i.e. technology, they lack something important, as Plato masterfully teaches us: "although man acquired in this way [through Hephaestus' fire and Athena's wisdom in the arts] the wisdom of daily life, civic wisdom he had not, since this was in the possession of Zeus" (Plat. *Prot.* 321 d). Actually, "Prometheus could not make so free as to enter the citadel which is the dwelling-place of Zeus, and moreover the guards of Zeus were terrible" (Plat. *Prot.* 321 d). Thus, something that still divides the divine from the human is "civic wisdom", which characterises the former, but not the latter. Speaking of "civic wisdom", which others translate as "political wisdom",[8] means speaking of "τὴν δὲ πολιτικὴν", which derives from one of the cornerstones of Western culture, i.e. the notion of πόλις (transliterated as *polis*), which underpins the notion of city not only by pointing at an aggregation of people and buildings but also, and especially, by pointing at a government model. More precisely, the kind of government model meant by the notion of πόλις implies the capacity to consider not only what is singular and individual but also what is plural and common when it comes to ruling. But humans, whose challenge has also to do with ruling, lack "political wisdom". If we ask ourselves what their lack may result in, then we may answer that, since they rule especially through technology, they seem to constitutively risk using technology without considering what is plural and common, specifically without giving it priority over what is singular and individual—humans seem to constitutively risk using technology for individual interests more than for common interests.

---

[8] See, for instance, the translation by Jowett, revised by Ostwald.

Surprisingly enough, the ancient Prometheus' last insight seems to describe our technological era even more than the past. And, if we move from the ancient Prometheus to the modern Prometheus, then his figure can illuminate the meaning of the current relationship between ourselves and technology even further. We should consider Mary Shelley's *Frankenstein. Or, the Modern Prometheus* as its key development (yet, without forgetting Percy Bysshe Shelley's, among others'). We should briefly recall that, according to Mary Shelley, Frankenstein is a talented and ambitious scientist who aims at challenging death itself, after his mother's death, as follows: first, he tries to create a human being from corpses, i.e. living human matter from nonliving matter (which is a divine prerogative), and, second, he tries to make his creature possibly resistant to death (which is a divine prerogative). Frankenstein succeeds, but the creature, who is a kind of deformed giant, terrifies him. The creature is also terrified by his reaction and runs away. The rest of the story is a series of tragedies for both Frankenstein and the creature. The former feels guilty and ends up being indirectly one of the creature's victims. The latter ends up killing after having been persecuted as a monster, even though he helped a family of farmers. A most interesting detail is the following: the creature makes Frankenstein an offer according to which, if the latter would have created an analogous woman, then the former would have disappeared with her far from humans. Frankenstein, who started creating the analogous woman, stops for fear that the creatures, with their possible offspring, would have been ever more terrifying for humans he wants to protect. Thus, the creature, first, punishes Frankenstein and, second, commits suicide by setting fire to himself to make sure that others could not understand and, consequently, replicate what started his life.

Before reflecting on the last most interesting detail, which distinguishes the modern Prometheus from the ancient Prometheus by making the former the latter's historical development, we should consider what they share. First, both the ancient Prometheus and Frankenstein aim at making the human closer to the divine. As far as the former is concerned, he gives humans Hephaestus' fire, Athena's wisdom in the arts and "an upturned face" and tells "them to lift their eyes to the stars". As far as the latter is concerned, he is described not only as trying to obtain the divine prerogatives of both creating living human matter from nonliving matter and making his creature possibly resistant to death but also as trying to obtain both a kind of divine omniscience ("It was the secrets of heaven and earth that I desired to learn", Shelley, 1994 [1818]: 36[9]) and a kind of divine omnipotence (he wants the "glory [that] would attend the discovery if I could banish disease from the human frame and render man invulnerable to any but violent death!", Shelley, 1994 [1818]: 37[10]). Even though both the ancient Prometheus and Frankenstein aim at making the human closer to the divine, there is an important difference between them: in general, divine

[9] See also the following words: "After days and nights of incredible labour and fatigue, I succeeded in discovering the cause of generation and life; nay, more, I became myself capable of bestowing animation upon lifeless matter" (Shelley, 1994 [1818]: 50).

[10] See also the following words: "this discovery was so great and overwhelming" (Shelley, 1994 [1818]: 50).

prerogatives increase and, in particular, Frankenstein's technology makes humans doubly creators, in that they create not only nonliving matter but also living human matter—the modern Prometheus' technology makes humans equal the divine, in that they are equally creators of living human matter.

Second, both the ancient Prometheus and Frankenstein express a notion of technology that implies, sometimes directly and sometimes indirectly, evil in addition to good. As far as the former is concerned, his myth results in Pandora's myth, as we have seen. As far as the latter is concerned, evil is expressed both in terms of his flaws ("As the minuteness of the parts formed a great hindrance to my speed, I resolved, contrary to my first intention, to make the being of a gigantic stature", Shelley, 1994 [1818]: 51. Finally, "disaster is irreparable", Shelley, 1994 [1818]: 71) and in terms of the creature's flaws, which are sometimes direct and sometimes indirect ("I was in reality the monster that I am, I was filled with the bitterest sensations of despondence and mortification. Alas! I did not yet entirely know the fatal effects of this miserable deformity", Shelley, 1994 [1818]: 109. Even though "I was benevolent and good [...] [,] misery made me a fiend", Shelley, 1994 [1818]: 96). Yet, even more interestingly, the solution to technology's flaws is still technology, from the ancient "*Elpis*", i.e. hope, which occupies the jar together with evil, to the modern creature's suicide committed, meaningfully, by setting fire to himself, i.e. by using technology's symbol itself to solve technology's flaws—technology is both the starting point (for instance, as what makes humans start being human after having stopped being "like swarming ants") and the ending point (for instance, as what has to set fire to itself, in that no humans solve it).

If we move from what the modern Prometheus shares with the ancient Prometheus to what distinguishes the former from the latter, then we should consider details that can make us reflect upon a trajectory closer to our technological era. I believe that the key issue to investigate the essential difference between ancient culture and modern culture and, consequently, the trajectory from the latter to contemporary culture is what is symbolically represented by the different kinds of punishments. More precisely, I believe that the different kinds of punishments characterising the ancient Prometheus and the modern Prometheus are illuminating when it comes to understanding the relationships between the technological, the human and the divine in terms of their crucial hierarchies, from power (rulers and ruled) to ontology (creators and created).

A promising starting point is offered by the following words, which Frankenstein's creature exclaims: "You are my creator, but I am your master: obey!" (Shelley, 1994 [1818]: 162). Clearly enough, a paradoxical overturning of both power hierarchies and ontological hierarchies is expressed. As far as power hierarchies are concerned, technology (symbolised by the scientist's creature) can rule ("obey!"). As far as ontological hierarchies are concerned, technology (symbolised by the scientist's creature) can move from something created (being the creature) to something that can create, at least potentially (being a "master", specifically "your master").

The different kinds of punishments characterising the ancient Prometheus and the modern Prometheus can illuminate even further. If we consider the ancient Prometheus, then the following order is expressed: there is one punisher, i.e. Zeus (the divine), and there are two punished, i.e. humans and Prometheus (the human and the relationship between humans and technology). It is worth noting that fire (technology itself) does not emerge: it is something that does not punish and is not punished. Conversely, if we consider the modern Prometheus, then the following order is expressed: there are three punishers, i.e. Frankenstein (the relationship between humans and technology), humans (the human) and the creature (technology), and there are three punished, i.e. Frankenstein (the relationship between humans and technology), humans (the human) and the creature (technology).[11] From a philosophical perspective, at least three issues are key. First, the divine as a player disappears, which may mean that it moves from being an ontological dimension in itself to being nothing but what other ontological dimensions aspire to obtain. Second, power hierarchies and ontological hierarchies between players disappear, in that Frankenstein, humans and the creature are both punishers and punished, which introduces the idea according to which moving from ancient culture to modern culture means moving from discernible differences (especially between what is human and what is technological) to indiscernible differences, which even disappear (especially between what is human and what is technological). And, as far as the divine is concerned, its modern destiny seems analogous: the more the human dissolves into the technological, the more the divine dissolves into the human—moving from ancient culture to modern culture and, as we shall see, moving from the latter to contemporary culture seems to mean the radical overturning of both power hierarchies and ontological hierarchies, in that what is inferior increasingly aspires to obtain what is superior and, finally, their differences end up disappearing.

More precisely, the move from ancient culture to modern culture is marked especially by the dissolution of the human into the technological and, as we shall see, the move from modern culture to contemporary culture is marked especially by the dissolution of the divine not only into the human but also into the technological.

There is a third key issue: technology as a player appears, in that the creature is both punisher and punished. Moreover, as we have seen, the creature punishes himself by setting fire, i.e. technology's symbol itself, to himself, which means that technology even seems to survive both the human and the divine (as contemporary culture shall make us reflect upon). Technology appears as a crucial player: it is the creature himself who can obtain (and not only make humans obtain) divine prerogatives, in that it is the creature himself who can be possibly resistant to death (and not only make humans challenge death by creating living human matter from nonliving matter). Thus, what the move from the ancient Prometheus to the modern Prometheus shows us is that, paradoxically enough, the historical development of the relationship between humans and technology, together with its trajectory, means technology's empowerment even more than humans' empowerment—technology goes even further than humans by moving from not being a player to being not

[11] As we have briefly recalled, both Frankenstein and humans punish the creature, who, finally, punishes both of them. Thus, both Frankenstein and humans are punished by the creature, who is initially punished by both of them.

only the human way to challenge the divine but also, and especially, something that challenges both the human and the divine.

The trajectory from modern culture to contemporary culture may be described as follows. First (the modern Prometheus), technology makes humans survive and even rule over nature, which means that, from a hierarchical perspective, the human can be superior to nature, but inferior to the divine, even though the former can potentially challenge the latter through its divine prerogatives (Hephaestus' fire and Athena's wisdom in the arts). Second (the modern Prometheus), technology makes humans create living human matter from nonliving matter and even challenge death, which means that, from a hierarchical perspective, the human can equal the divine: the former can actually challenge the latter through its divine prerogatives (creating living human matter from nonliving matter and even challenging death through the creature's resistance to death). The modern Prometheus' move has to do with a more sophisticated technology that makes the human as creator as the divine—and a more sophisticated technology also means that technology itself starts obtaining divine prerogatives (the creature's resistance to death).

1.2 Human Creation of a Technological Divine

What is the contemporary Prometheus' move? Trying to read one's own era is always complicated. Yet, as I have tried to do elsewhere,[12] considering artistic representations may be a promising way to get insights. On the following pages, I shall briefly focus on at least one most promising artistic representation, i.e. a novel, aiming at better understanding our technological era, and I shall argue that there are meaningful analogies between the novel, the figure of Prometheus (the novel clearly expresses humans' attempt to obtain divine prerogatives through technology) and, especially, our technological era itself (the novel clearly expresses that technology, specifically digital technology, is precisely what finally obtains divine prerogatives through humans' attempt).

The novel is Eggers' *The Circle. A Novel*, and the reasons why I choose it are not only its exceptional capacity to represent, in particular, humans' attempt to obtain divine prerogatives through technology and technology as what finally obtains divine prerogatives through humans' attempt but also its exceptional capacity to represent, in general, the close relationship between the three ontological dimensions of the human, the technological and the divine on which, as we have seen, we can find fundamental insights even in the cradle of Western culture. In the novel, we read the following words pronounced by a mysterious character, who, interestingly enough, studied divinity, worked as a computer scientist and ended up falling into alcoholism: "Now we're all God. Every one of us will soon be able to see, and cast judgment upon, every other. We'll see what He sees. We'll articulate His judgment. We'll channel His wrath and deliver His forgiveness. On a constant and global level. All

[12] Especially in Chiodo (2020a).

religion has been waiting for this" (Eggers, 2014: 398). The mysterious character talks about the novel technology developed by the three creators, i.e. a kind of divine trinity, of the technology company aiming at what follows, as one of them tells. First, "We will become all-seeing" (Eggers, 2014: 71) through most sophisticated technologies allowing "Full transparency [that] would bring full access" (Eggers, 2014: 470). Second, "We will become [...] all-knowing" (Eggers, 2014: 71) through most sophisticated technologies allowing "no more not-knowing" (Eggers, 2014: 470): "I had the idea that I would endeavour, in my lifetime, to know every person on Earth. Every person, even if just a little bit. To shake their hand or say hello. And when I had this inspiration, I really thought I could do it. Can you feel the appeal of a notion like that?" (Eggers, 2014: 469). Third, most sophisticated technologies will allow "the perfectibility of human beings. [...] I think we can be perfect or near to it. And when we become our best selves, the possibilities are endless. We can solve any problem. We can cure any disease, end hunger, everything, because we won't be dragged down by all our weaknesses, our petty secrets, our hoarding of information and knowledge. We will finally realize our potential" (Eggers 2014: 294). From a philosophical perspective, several issues are key in addition to the meaningful analogy between the three creators (of novel technologies and, consequently, of novel humans) and the divine trinity. Clearly enough, the first point ("We will become all-seeing") correlates with an idea of technological omnipresence, the second point ("We will become [...] all-knowing") correlates with an idea of technological omniscience and the third point ("the perfectibility of human beings", who "can solve any problem") correlates with the idea of technological omnipotence. It is worth noting that, again, the most typical divine prerogatives of omnipresence, omniscience and omnipotence are obtained not only by humans but also by technologies themselves, especially in the case of omnipresence (strictly speaking, technologies, and not humans, are omnipresent) and in the case of omniscience (strictly speaking, technologies, and not humans, are omniscient).

Thus, the contemporary Prometheus' move seems to emerges as follows: technology makes humans obtain the most typical divine prerogatives of omnipresence, omniscience and omnipotence, which seems to mean that, from a hierarchical perspective, the human can equal, again, the divine. Yet, I believe that the contemporary Prometheus' move emerges in a more disruptive way. On the one hand, humans equal the divine by obtaining omnipresence, omniscience and omnipotence. On the other hand, something more disruptive happens: humans create a kind of technology that itself is divine by obtaining omnipresence, omniscience and omnipotence—which means that humans themselves create a kind of divine and, consequently, surpass the divine itself.

When we have considered the modern Prometheus, we have reflected upon humans as doubly creators, in that technology makes them create not only nonliving matter (artefacts) but also living human matter (the creature). If we move from the modern Prometheus to the contemporary Prometheus, then we may reflect upon humans as triply creators, in that they can also create technology as a special kind of divine: on the one hand, it is an artefact, i.e. something immanent, but, on the other hand, it itself is omnipresent, omniscient and omnipotent, which are the most typical divine prerogatives of the divine as something transcendent.

If we think of current technologies we continuously use, from our computers to our smartphones, especially when we google something (it is no coincidence that a technology company's name has quickly become a verb), then the phenomenon described above clearly emerges. Whenever we google something (for instance, the symptom we experience), we contribute to create, through our data, a special kind of entity: on the one hand, it is an artefact, i.e. something immanent, but, on the other hand, it itself is omnipresent (by being everywhere: it is always with us, from our desks to our pockets), omniscient (by knowing everything: it knows not only the answers to our questions but also ourselves by tracking us) and omnipotent (by having more and more power: it has more and more power over us, from shaping our worldviews to shaping our decisions and, consequently, our actions). In addition to Google's algorithm, we may think of several analogous cases in which we contribute to create, through our data, a special kind of entity—more precisely, a technological divine, which is immanent, i.e. one of our artefactual creations, and, at the same time, divine by increasingly obtaining omnipresence, omniscience and omnipotence. For instance, we may also think of the following cases, in which omnipotence plays a disruptive role because of biases (see at least Noble, 2018 and Tobias, 2019, and even Weizenbaum, 1976) that, even unconsciously, characterise algorithms trained to include specific sets of data (for instance, data related to whites, Caucasians and men) and, consequently, to exclude specific sets of data (for instance, data related to blacks, Asians and women): Hewlett-Packard's algorithm, which ended up discriminating against blacks when using webcams (see Leonard, 2009), Nikon's algorithm, which ended up discriminating against Asians when using cameras (see Rose, 2010), and Amazon's algorithm, which ended up discriminating against women when recruiting staff (see Dastin, 2018).

What distinguishes the contemporary Prometheus from the modern Prometheus is not that humans' creation is capable of mastering its creator, in that both the creature and, for instance, Google are capable of mastering humans (even though the former exclaims that "You are my creator, but I am your master: obey!" and the latter masters silently). Conversely, what distinguishes contemporary culture from modern culture is that technology masters analogously to the divine. Frankenstein's creature is not omnipresent, omniscient and omnipotent: from an ontological perspective, he is a kind of super-human, who can be possibly resistant to death, but who ends up committing suicide. Conversely, our creatures as the information technologies described above master analogously to the divine, in that, from an ontological perspective, they move from being a kind of super-human to being a kind of divine—an immanent divine, but omnipresent, omniscient and omnipotent.

Moreover, we may identify a further divine prerogative obtained by the information technologies described above, i.e. inscrutability—in addition to omnipresence, omniscience and omnipotence, current technologies also seem to obtain divine inscrutability. As several authors stress,[13] algorithms as the basis of a great number of current technologies can be defined as black boxes, in that the reasons why a given input results in a given output can be inscrutable. If we reflect upon the meaning of inscrutability from a philosophical perspective, then we should consider at least two issues. First, algorithms' inscrutability makes them a kind of divine not only in that their reasons are inscrutable but also, and especially, in that the way in which humans acknowledge their outputs is a kind of act of faith (as I have argued elsewhere through a real example,[14] if you happen to receive a letter from a hospital in which it is literally written that, according to the "predictive algorithm" that processed the data from the questionnaire you filled in, you are not eligible to take a specific genetic test, in that your risk of being genetically predisposed to suffer from a specific cancer is not high enough, then you can do nothing but acknowledge the "predictive algorithm['s]" output through a kind of act of faith, especially if, as it is the case, you do not know what an algorithm is, but the adjective "predictive" reassures you and no doctor tells you what "predictive algorithm" means). More generally, speaking of acknowledging algorithms' outputs through a kind of act of faith means speaking of a key issue of our technological era: the belief that computation's outputs are not only reliable but also more reliable than the human mind's outputs (see at least Porter, 1995. I shall develop the issue of the increasing faith in computation in Chap. 4). Second, algorithms' inscrutability also makes them a kind of divine by increasing their omnipotence over humans. If you know one's reasons, then you can understand them and possibly argue against them. But, if you do not know one's reasons, then, first, you cannot understand them and possibly argue against them and, second, you may easily happen to fall into a kind of automatism that may be analogous to a kind of manipulation: the more you experience that you cannot understand something and argue against something, the less you exercise your critical thinking when you have to do with decisions resulting from algorithms (in my real example, it is no coincidence that speaking of a "predictive algorithm" means speaking of a divine prerogative, i.e. the capacity to predict, as we have seen: if you are increasingly told that the most reliable decisions result from something that has the capacity you do not have, then you increasingly stop thinking and, consequently, start automating your actions, as I shall argue in the following chapters).

Articulations of the general issue of the human creation of a technological divine shall also be developed through cases in point analysed both in Chap. 2 (autonomous technologies) and in Chap. 3 (self-tracking technologies and the quantified self). But I believe that a further philosophical perspective should be added before proceeding, in that it may be a promising way to read our era in general and our technological era in particular—I believe that our technological era may be read as the most radical form of anarchism we have ever experienced in Western culture, as I myself

---

[13] See at least Finn (2017) for a philosophical perspective on the issue of algorithms' inscrutability. It is no coincidence that the issue of algorithms' explainability is key in the current debate (see at least Wallach & Allen, 2008; Jobin, Ienca & Vayena, 2019; Miller, 2019; Rosenfeld & Richardson, 2019; Barredo Arrieta et al., 2020; Coeckelbergh, 2020b; Confalonieri et al., 2021 and Palmirani & Sapienza, 2021).

[14] Especially in Chiodo (2020a), Chap. 3, and (2020b).

extensively argued elsewhere.[15] We have seen that the move from modern culture to contemporary culture is marked especially by the dissolution of the divine not only into the human but also into the technological. And we have also seen that dissolving the divine into the technological means moving from a transcendent divine (as it has been thought of by Western culture for millennia) to an immanent divine (as I argue by reading our technological era from a philosophical perspective). A further logical step is the following: even from an etymological perspective, speaking of an immanent divine means speaking of a radically anarchic move: "anarchism" as the radicalisation of "anarchy" means radical absence (ἄν, transliterated as *an*) of something that rules (ἄρχω, transliterated as *archo*)—speaking of anarchism means speaking of radical absence of a ruler, i.e. radical rulerlessness. In millennia of Western culture, the ruler is especially the divine, in that it is thought of as a superior ontology from both a chronological perspective (according to which the divine is the ruler as the beginning of everything else) and a logical perspective (according to which the divine is the ruler as the cause of everything else). As we have seen, modern culture starts a phenomenon that contemporary culture radicalises: the superior ontology increasingly dissolves into both the human and the technological—consequently, the ruler increasingly disappears. We may translate the case of the creature who masters his creator and, especially, the case of Google that masters humans (in addition to my real example of the equally mastering "predictive algorithm") into the following philosophical words: the ruler increasingly moves from transcendence to immanence. More precisely, the ruler increasingly moves from an ontological dimension irreducible to that of what it rules to an ontological dimension reducible to that of what it rules—which means crucial epistemological and practical consequences readable as anarchism.

The cases in point analysed in the following chapters shall show the correlation between the ontological overturning described above and anarchism. Yet, the issue may be introduced by my real example. If you happen to receive the letter described above, and no doctor tells you, who are a nonexpert, what "predictive algorithm" means, then you are likely to act as follows. First, you google the words "predictive algorithm", in that your smartphone is in your pocket as a kind of ready-to-use omniscient technology. Second, you read the first thing that appears at the top of the list of links: "Predictive algorithms use one of two things: machine learning or deep learning. Both are subsets of artificial intelligence (AI). [...] Random Forest: this algorithm is derived from a combination of decision trees, none of which are related, and can use both classification and regression to classify vast amounts of data".[16] Third, you have two options: on the one hand, you may decide to google at least the words "machine learning", "deep learning" and "artificial intelligence" and, on the other hand, you may decide to rely on the reassuring words added to the equally reassuring adjective "predictive", i.e. "learning", "deep learning", "intelligence" and "vast amounts of data". You are likely to choose the second option, in that, in addition to several reassuring words, you end up telling yourself that the meanings of the words

[15] Especially in Chiodo (2020a).

[16] I quote what I found when I googled the words "predictive algorithm" in September 2021.

"machine learning", "deep learning" and "artificial intelligence" may turn out to be equally incomprehensible and, especially, you are in a hurry.

If we reflect upon the epistemological meaning of the scenario described above, then we should say that it shows a kind of epistemological surrender—more precisely, a kind of crisis of expertise readable as epistemological anarchism. The expert as the mediator between knowledge and the nonexpert disappears. On the one hand, there is no doctor explaining. On the other hand, there is Google explaining, which means that there is no reliable knowledge: even though the definition of "predictive algorithm" is written by experts, reliable knowledge requires the expert's capacity to correctly subsume the particular to the universal, as Kant masterfully teaches us (see Kant, 1781), which means that reliable knowledge requires the expert's capacity to correctly subsume the "predictive algorithm" of the letter described above to the definition of "predictive algorithm". Conversely, you, as a nonexpert, are not likely to obtain reliable knowledge (which is the reason why it is recommended not trying to self-diagnose by googling our symptoms: expertise is not a matter of data, but a matter of the capacity to correctly read data. Otherwise, we would obtain the worst epistemological outcome, as we learn even from Socrates: not knowing that we do not know anything and, consequently, taking what is false instead of what is true).

And taking what is false instead of what is true is an epistemological scenario that leads us to a practical scenario—more precisely, a practical scenario characterised by a kind of randomness readable as practical anarchism: the first thing that appears at the top of the list of links today may change tomorrow for reasons that may have nothing to do with reliable knowledge. As I shall develop in the following chapters, the increasing disappearance of the expert as the mediator between knowledge and the nonexpert, i.e. the increasing replacement of the former with a kind of ready-to-use omniscient technology, leads us to an equally increasing randomness when, finally, we act, in that our actions result not only from unreliable knowledge but also from a series of random circumstances: when we google the words "predictive algorithm", we are likely to follow the first thing that appears today (which may change tomorrow for reasons that may have nothing to do with reliable knowledge) and, when we decide what option to choose, we are likely to follow the idiosyncratic circumstances that affect us today, i.e. needing reassuring words and being in a hurry (which may change tomorrow). Finally, and more dramatically, we are likely to confuse the output of the "predictive algorithm", i.e. a risk that is not high enough, with a low risk, which may discourage us from preventing severe health problems that may happen.

The key reason why the scenarios described above are readable as anarchism is that, both in the epistemological case and in the practical case, something that actually rules (ἄρχω) is radically absent (ἄν). Clearly enough, speaking of actually ruling means speaking of something exceedingly different from a mere relationship of cause and effect. The kind of ruling meant by the Greek word ἄρχω is not a matter of being a random cause of an effect, as it is the case of unreliable knowledge causing a series of decisions and actions. Conversely, it is a matter of authority[17]: ἀρχός (transliterated as *archos*) is "a leader, chief, commander"[18]—which means that ruling requires the capacity to rule, i.e. to know what to do and to lead to it (and not merely to cause an effect). And, again, if there is no ruler, i.e. knower and leader, as it is the case of the doctor's expertise replaced by the randomness of the nonexpert's googling, then there is anarchism: our decisions (from an epistemological perspective) and our actions (from a practical perspective) are underpinned by nothing but the randomness of something that causes without ruling.

Thus, the human creation of a technological divine may be thought of as a radically anarchic move for at least two key reasons: first, humans succeed in creating a technological divine that is totally immanent and, second, the totally immanent technological divine does not succeed in being an actual ruler in itself—actually ruling seems to require actual expertise and actual expertise seems to require a hierarchical difference from a totally immanent technological randomness.

1.3 Radical Overturnings

The historical development of the figure of Prometheus may be read as a series of ontological overturnings. We have identified overturned power hierarchies, in that something ruled becomes something that can rule: in ancient culture, humans become capable of ruling nature, in modern culture, technology becomes capable of ruling humans and, in contemporary culture, both humans and technology become capable of ruling the divine in the sense that they overturn it (as the ancient Prometheus' prophecy predicts: "until Zeus is hurled from his sovereignty"). And we have identified overturned ontological hierarchies, in that something created becomes something that can create: in ancient culture, humans become capable of creating technologies as nonliving matter (artefacts), in modern culture, humans become capable of creating technologies as living human matter (the creature) and, in contemporary culture, humans become capable of creating the divine in the sense that they make omnipresent, omniscient, omnipotent and inscrutable technologies. Thus, the trajectory is clear: speaking of overturnings means speaking of moves of what is somehow inferior to what is somehow superior. Finally, their differences end up disappearing.

Several kinds of differences end up disappearing: between rulers and ruled, between creators and created and, especially, between transcendence and immanence. As we have seen by considering epistemological and practical consequences, we should add other kinds of differences that disappear. More precisely, we have reflected

[17] Middle Liddell, see http://www.perseus.tufts.edu/hopper/morph?l=a%29%2Frxw&la=greek&can=a%29%2Frxw0&prior=h(#Perseus:text:1999.04.0058:entry=a)/rxw-contents (accessed in September 2021).

[18] Middle Liddell, see http://www.perseus.tufts.edu/hopper/morph?l=a%29rxo%2Fs&la=greek&can=a%29rxo%2Fs0#lexicon (accessed in September 2021).

upon expertise dissolving into randomness. We should add that the increasing disappearance of the difference between expertise and randomness means something specific in Western culture: the equally increasing disappearance between what is universal (as a rule given by an expert) and what is particular (as an idiosyncratic circumstance given by randomness)—which also implies the equally increasing disappearance between the realm of ideality as the dimension of universal rules and the realm of reality as the dimension of particular idiosyncratic circumstances (see at least Rescher, 1987; Brzeziński et al., 1990; Emmet, 1994; Berlin, 1996 and Jullien, 2009).

Our technological era offers a great number of examples to investigate the issue of differences that disappear between universality and particularity, specifically between expertise as rulers and randomness as idiosyncratic circumstances. For instance, whenever we use a digital platform to try to move from representative democracy (in which experts work on the drafting of laws) to direct democracy (in which nonexperts also work on the drafting of laws), as it is the case of the Italian Five Star Movement (see at least Santoro, 2012; Corbetta & Gualmini, 2013; Tronconi, 2015; Iacoboni, 2018 and Biondo & Canestrari, 2019), we replace expertise as the capacity to rule, i.e. to know what to do and to lead to it, with randomness as merely causing an effect. It is no coincidence that, according to the website of the Five Star Movement, the move from representative democracy to direct democracy means the removal of "the mediation of governing or representative bodies, recognising to all citizens the governing and steering role normally attributed to few".[19] Yet, the removal of "the mediation" characterising our technological era frequently results in the removal of expertise: when we "recognis[e] to all citizens the governing and steering role normally attributed to few", we entrust the drafting of laws (which are by definition universal, i.e. underpinned by the capacity to move from what is right for me as an individual to what is right for us as a community) to nonexperts who may even happen not to know that the move from considering the particular to considering the universal is crucial in the drafting of laws.

Several current technologies are characterised by the possibility, which is even their precise aim (as we shall see especially in the case of self-tracking technologies and the quantified self), of making hierarchical differences disappear. Yet, the meaning of ἄρχω implies hierarchical differences, without which we fall into anarchism—what I shall argue in the following chapters is precisely that several current technologies mark our era as a radical overturning of hierarchical differences, starting with the fundamental dissolution, first, of the human into the technological and, second, of the divine not only into the human but also into the technological (which is the key reason why our technological era may be read as the most radical form of anarchism we have ever experienced in Western culture).

Before proceeding, we should add one last key issue. If it is true that what rules has been especially identified for millennia, specifically between ancient culture and modern culture, with the divine, i.e. something heteronomous, it is also true that the philosophy of the last three centuries has also systematically introduced the

[19] See https://www.movimento5stelle.it/ (accessed in September 2021, my translation).

idea according to which what rules can, and even should, be autonomous. From an etymological perspective, the adjectives heteronomous and autonomous share precisely the Greek word νόμος (transliterated as *nomos*), i.e. "law", which strictly correlates with ruling. But, on the one hand, the adjective heteronomous identifies a ruler that is "other" (ἕτερος, transliterated as *heteros*), i.e. not coinciding with what is ruled, and, on the other hand, the adjective autonomous identifies a ruler that is "itself" (αὐτός, transliterated as *autos*), i.e. coinciding with what is ruled—speaking of autonomy means speaking of the coincidence between ruler and ruled, in that the rule is self-given.

Thus, the most critical question we should ask ourselves is the following: if ruling itself experiences a crisis, then what is the destiny of autonomy in our technological era? More precisely, what is the destiny of human autonomy in our technological era?

There is a sense in which the dissolution of the divine into the human correlates with the idea according to which human autonomy underpins human identity and dignity. Kant's philosophical work on the notion of autonomy is fundamental[20]: "Autonomy of the will is the property of the will by which it is a law to itself" (Kant, 1785: 4: 440), and being "a law to itself" is "the ground of the dignity of human nature and of every rational nature" (Kant, 1785: 4: 436), in that, as Korsgaard writes, one's autonomous will and, consequently, one's autonomous law, i.e. rule, "express your identity, your nature" (Korsgaard, 1996b: 84). More precisely, one's autonomous laws, i.e. rules, result from "the conceptions of ourselves that are most important to us [...]. For to violate them [one's autonomous laws, i.e. rules] is to lose your integrity and so your identity, and no longer to be who you are. That is, it is no longer to be able to think of yourself under the description under which you value yourself and find your life worth living and your actions worth undertaking" (Korsgaard, 1996b: 84). Kant's philosophical work on the notion of autonomy is fundamental to make contemporary philosophers further stress the key role played by autonomy in underpinning not only human identity and dignity but also human freedom and morality as specifications of what human identity and dignity actually mean: human "freedom and the will's own lawgiving are both autonomy" (Kant, 1785: 4: 450) and "when we think of ourselves as free we transfer ourselves into the world of understanding as members of it and cognize autonomy of the will along with its consequence, morality" (Kant, 1785: 4: 453). We have reflected upon something analogous when we have argued that speaking of actually ruling means speaking of something exceedingly different from a mere relationship of cause and effect. The

[20] As far as the perspective I consider is concerned, see at least Hill (1989, 2000), Korsgaard (1996a, 1996b), Guyer (2003), Taylor (2005b), Reath (2006), Deligiorgi (2012) and Sensen (2013). As far as the notion of autonomy more generally considered is concerned, see at least Berlin (1958), Frankfurt (1971, 1988 and 1999), Rawls (1971), Haworth (1986), Nagel (1986), Raz (1986), Dworkin (1988), Christman (1989, 1991, 2001 and 2005), Meyers (1989), Mele (1991, 1995), Ekstrom (1993), Benson (1994), May (1994a), Berofsky (1995), Lehrer (1997), Fischer and Ravizza (1998), Oshana (1998), Schneewind (1998), Cuypers (2001), Frankel Paul et al. (2003), Friedman (2003), Taylor (2005a), Darwall (2006), Larmore (2008) and Killmister (2017), in addition to Dryden available at https://iep.utm.edu/autonomy/.

effect of a cause does not necessarily require freedom and morality, in that it may be nothing but the output of an automatism. Conversely, ruling necessarily requires freedom and morality if it is true that it means, as we have seen, the capacity to rule, i.e. to know what to do and to lead to it (and not merely to cause an effect). Knowing what to do requires freedom for at least one reason, i.e. the possibility of reflecting upon and choosing between options (which is exceedingly different from the automatism between one cause and one effect). And leading to it requires morality for at least one reason, i.e. the responsibility taken and for which to be accountable (which is exceedingly different from the neutrality between one cause and one effect). Thus, speaking of ruling means speaking of at least two prerogatives: on the one hand, reflecting upon and choosing between options and, on the other hand, taking responsibility and being accountable for it—speaking of ruling means speaking of a rule that can be demanded of others only if it is "a law to [one]self": being autonomous means being free not by requiring loss of rules, as it is the case of anarchism, but by requiring self-given rules, which must be there ("the will must have a law, but because the will is free, it must be its own law. And nothing determines what that law must be. All that is has to be is a law", Korsgaard, 1996b: 81).

Thus, again, if our technological era is characterised by an increasing loss of rules, which are increasingly replaced by a kind of automated randomness that may be read as the most radical form of anarchism we have ever experienced in Western culture, then what is the destiny of human autonomy precisely as what requires rules, i.e. self-given rules, from its etymology to philosophy? More precisely, what is the destiny of human "self" (αὐτός) if human "law" (νόμος), i.e. rule, increasingly dissolves into a kind of automated randomness, specifically technology's automated randomness?

The question becomes even more critical when we consider the history of Western culture, specifically the history of Western philosophy. If it is true that the idea according to which what rules can, and even should, be autonomous has been systematically introduced by the philosophy of the last three centuries, especially from Kant onwards, it is also true that the condition of possibility for thinking of autonomy as what underpins human identity and dignity is ancient—which means that our technological era actually challenges one of the most essential cornerstones of the definition itself of human identity and dignity in Western culture.

It is worth briefly recalling that we can find an idea of autonomy as a kind of rational self-determination leading to moral actions in Plato's philosophy, in that human rationality is thought of as ruling over human irrationality. The former is described as what is actually human and the latter is described by referring to animals: "'all our actions and words should tend to give the man within us complete domination over the entire man and make him take charge of the many-headed beast—like a farmer who cherishes and trains the cultivated plants but checks the growth of the wild—and he will make an ally of the lion's nature, and caring for all the beasts alike will first make them friendly to one another and to himself, and so foster their growth'. 'Yes, that in turn is precisely the meaning of the man who commends justice'" (Plat. *Resp.* 9, 589 a, b). We can also find an idea of autonomy as a kind of rational self-determination leading to moral actions in Aristotle's philosophy, in that the former, specifically human rationality, is thought of not only as what defines humans' essence but also as the condition of possibility for the latter: the good human being "work[s] out the good [...], and does so for his own sake (for he does it for the sake of the intellectual element in him, which is thought to be the man himself); and he wishes himself to live and be preserved, and especially the element by virtue of which he thinks. [...] and the element that thinks would seem to be the individual man, or to be so more than any other element in him" (Arist. *Eth. Nic.* 9, 1166 a 17–23). As we have seen, modern philosophy plays a key role in thinking of autonomy as what underpins human identity and dignity. In addition to Kant's philosophical work, Rousseau also contributes to the reflection upon autonomy by correlating self-determination with self-development (see especially Rousseau 1762). In the following century, Mill correlates self-determination with the notion of individual, whose importance further increases. Moreover, in his words we can find Kant's hierarchical difference between what is autonomous and what is heteronomous: "A person whose desires and impulses are [...] not his own has no character, no more than a steam engine has a character" (Mill, 1859: 73). It is worth noting that, interestingly enough, we can also find the idea according to which, if the reasons why one acts are not autonomous ("desires and impulses are [...] not his own"), then one is somehow reduced to something automated ("a steam engine"). Finally, in contemporary philosophy, autonomy continues to be thought of as playing a key role in underpinning human identity and dignity, and both ancient philosophy and, especially, Kant's philosophical work are fundamental reference points (see note 20). In addition, contemporary philosophers also work on autonomy as a kind of individualism, especially procedural individualism (see especially Dworkin, 1988; Frankfurt, 1988; Ekstrom, 1993; Friedman, 2003 and Bratman, 2007). Thus, from ancient philosophy to contemporary philosophy, especially in the last three centuries, autonomy is one of the most essential cornerstones of the definition itself of human identity and dignity in Western culture—moreover, as we have seen by briefly recalling fundamental authors of the history of Western philosophy, autonomy is a privileged way to strictly correlate what humans are, i.e. capable of self-giving rules, with their morality in general and with their moral actions in particular, i.e. "commend[ing] justice", "work[ing] out the good" and "ha[ving] a character".

The focus on the key role played by autonomy aims at clarifying the sense in which I correlate technology, on the one hand, with the overturning of human autonomy, on the other hand. I shall develop the issue of autonomy in the following chapters, but I have briefly anticipated its milestones to clarify the direction of my work as follows. Our technological era increasingly shows an immanentisation of the ruler. The phenomenon described above frequently means an immanentisation of the divine as a radically anarchic move: the divine, i.e. its most typical prerogatives, moves from transcendence to technological immanence. A crucial consequence is that, from humans' perspective, the ruler's heteronomy equally moves from belonging to transcendence to belonging to technological immanence—which somehow sabotages the ruler itself (as we shall see in the following chapters), in that it is a human creation. We may imagine a kind of circle: if it is true that the ruler as technology can master humans, it is also true that the latter can also master the former in a way that is impossible if the ruler is thought of as transcendent.

But the phenomenon of the immanentisation of the ruler also means something that has to do with the relationship between the human and the technological, and not the divine—more precisely, it also means another kind of anarchic move: from humans' perspective, the loss of the ruler's autonomy. Again, the ruler itself is sabotaged, but in another way. Whenever we make an immediate decision based on an algorithm's output, we lose the kind of mediation that somehow identifies the ruler's autonomy, specifically its capacity to give itself its own rules. The reason why mediation somehow identifies the ruler's autonomy is that it is precisely technology's immediacy what may sabotage our exercise of critical thinking both when we reflect upon the reasons underpinning the rule we give ourselves and when we reflect upon the reasons underpinning the expert's rule we, finally, give ourselves. Even though the difference between the dimension of what rules and the dimension of what is ruled is not the difference between transcendence and immanence, there is a crucial difference: as Kant masterfully teaches us, when we give ourselves something that is an actual rule, we make it move from the realm of particularity to the realm of universality, in that we reflect upon its consequences as a rule applied by us and, consequently, by others with whom we interact—which means that, if we actually exercise our autonomy, then there is a crucial difference between the dimension of what rules and the dimension of what is ruled, in that the former somehow moves from the latter's particularity to universality. Conversely, when we make an immediate decision based on an algorithm's output, we sabotage the rule's possibility of moving from particularity to universality (we may use Kant's words and say that, if it is the case, then we do not "act as if the maxim of […] [our] action were to become by […] [our] will a universal law", Kant, 1785: 4: 421). For instance, I myself as an academic may say that we increasingly use algorithms' outputs, specifically rankings of publishers and journals, among others, to make immediate decisions on the quality of something that ends up correlating with the quality, and even with the career, of someone. I myself participated in boards in which we risked totally replacing the mediation implied by exercising our critical thinking with the immediacy implied by using rankings (as if exceptions, which are, for instance, articles whose quality does not coincide with journal rankings, do not exist). Yet, several authors recommend not relying on the immediacy of algorithms' outputs (see at least O'Neil, 2016). As far as rankings of publishers and journals are concerned, we can read in the San Francisco Declaration on Research Assessment (DORA): "Do not use journal-based metrics, such as Journal Impact Factors, as a surrogate measure of the quality of individual research articles, to assess an individual scientist's contributions, or in hiring, promotion, or funding decisions"[21] (in addition, see at least Singh Chawla, 2019 on manipulating metrics). But the increasing replacement of critical thinking's mediation with technology's immediacy seems unstoppable.

[21] See https://sfdora.org/read/ (accessed in September 2021).

And speaking of technology's immediacy means speaking of the move from autonomy to automation, in which, as I shall argue in Chap. 2, self-given rules disappear also from an etymological perspective. Thus, again, what does it mean to lose human autonomy as what has underpinned human identity and dignity for millennia, especially in the last three centuries?

Chapter 2
First Case in Point: Autonomous Technologies

2.1 Making Technologies Autonomous, Making Humans Automated

Surprisingly enough, if we pay attention to the current way we use language, then we realise that we increasingly shift the word "autonomy" from defining ourselves to defining technologies. As we have seen, especially in the last three centuries, we have thought of autonomy as "the property of the will by which it is a law to itself"—which means that autonomy is "the ground of the dignity of human nature" and, consequently, what "express[es] your identity, your nature". Yet, if we pay attention to the current way we use language, then we may easily happen to find the word "autonomy", specifically its adjective, as follows: autonomous systems, autonomous software, autonomous devices, autonomous applications, autonomous machines, autonomous vehicles, autonomous equipment, autonomous drones, autonomous weapons, autonomous robots, autonomous agents, autonomous workloads and so forth. From both an etymological perspective and a philosophical perspective, the shift described above is exceedingly meaningful. More precisely, I shall argue that it is a meaningful symptom of the radical overturnings characterising our era in general and our technological era in particular.

If we pay attention, for instance, to the case of autonomous vehicles, then we may easily happen to find the word "autonomy" as follows: "Today's automotive industry relies on three types of vehicle testing: via computer simulation, on real-world public roads, or behind closed doors at a private test track. Approaches vary, but a combination of all three approaches is deemed vital in order to safely introduce highly autonomous vehicles".[1] Interestingly enough, the quote starts and ends with words sharing the first half of their etymology: "automotive" and "autonomous", which share αὐτός as "itself". But they do not share the second half of their etymology. As

[1] See https://www.automotiveworld.com/articles/private-test-tracks-are-where-autonomous-vehicles-drive-risk-free/ (accessed in January 2021).

S. Chiodo, *Technology and the Overturning of Human Autonomy*,
Studies in Applied Philosophy, Epistemology and Rational Ethics 66,
https://doi.org/10.1007/978-3-031-26159-6_2

far as "automotive" is concerned, the etymological root is, first, the Greek ἀμείβω (transliterated as *ameibo*) as "changing" and, second, the Latin *moveo* as "moving". As far as "autonomous" is concerned, as we have seen, the etymological root is the Greek νόμος as "law". Thus, in the case of "automotive", we speak of something characterised by "self-motion" and, in the case of "autonomous", as we have seen, we speak of something characterised by "self-given law" (more precisely, the quote even speaks of "highly" "self-given law"). Both the expression "self-motion" and the expression "self-given law" describe things: the former is the adjective of "industry" and the latter is the adjective of "vehicles". Not surprisingly, "industry" correlates with "automotive", i.e. "self-motion": we can actually find "industries" of "self-moving" things, specifically vehicles. But it is quite surprising that "vehicles" correlate with "autonomous", i.e. "self-given law": can we actually find "vehicles" characterised by the capacity to "give themselves laws"? The provocative question may have an easy affirmative answer: we can actually find "vehicles" characterised by the capacity to "give themselves laws" in the sense that they can be based on machine learning algorithms. Yet, at least from a philosophical perspective, the provocative question cannot be answered by ignoring what the capacity to "give themselves laws" as the capacity to be "autonomous" actually means.

Several authors work on autonomy as something possessed by technology (see at least Winner, 1977). In what follows, I shall work on something quite different: autonomy as something possessed by humans and, consequently, the philosophical meaning of the shift from humans' autonomy to technology's autonomy, on the one hand, and the philosophical meaning of the shift from technology's automation to humans' automation, on the other hand.

Most interestingly, Kant himself refers to the notion of automation, at least from an etymological perspective, when he distinguishes between "*automaton materiale*, when the machine is driven by matter" (Kant, 1788: 5, 97), and "*automaton*" "*spirituale*, when it is driven by representations" (Kant, 1788: 5, 97). The case of "*automaton*" "*spirituale*" has to do with humans as follows: "if the freedom of our will were none other than [...] psychological and comparative but not also transcendental, i.e. absolute [...], then it would at bottom be nothing better than the freedom of a turnspit, which, when once it is wound up, also accomplishes its movements of itself" (Kant, 1788: 5, 97)—if we replace the kind of autonomy meant by something universal, i.e. rules, with the kind of "*automaton*" meant by something particular, i.e. idiosyncratic circumstances ("psychological and comparative but not also transcendental"), then we end up being nothing but "turnspit[s]", which means that our freedom ends up being nothing but randomness as the automatism between one cause and one effect (which, as we have seen, is exceedingly different from the capacity to rule). It is no coincidence that, if humans are not autonomous, i.e. are "turnspit[s]", then, according to Kant, the difference between them and things is nothing but a matter of adjectives ("*spirituale*" characterises humans and "*materiale*" characterises things), and not a matter of nouns, in that humans and things share "*automaton*".

Moreover, if humans replace autonomy with "*automaton*", then not only their freedom is undermined (it is "nothing better than the freedom of a turnspit") but also their morality is even more severely undermined ("no moral law is possible and no imputation in accordance with it", Kant, 1788: 5, 97). The relationship between the destiny of human autonomy and the destiny of human morality is crucial to understand our technological era: if it is true that our technological era may be read as shifts, first, from humans' autonomy to technology's autonomy and, second, from technology's automation to humans' automation, it is also true that we should question the relationship between humans' decreasing autonomy and increasing automation, on the one hand, and human morality, on the other hand. For instance, we may imagine to use "highly autonomous vehicles": in the case of a fatal crash,[2] who is morally responsible, if any? We as users? Engineers as designers? Anyone else? The current debate on the issue (see at least Matthias, 2004; Goodall, 2014; Hevelke & Nida-Rümelin, 2015; Coeckelbergh, 2016; Santoni de Sio, 2017; Nyholm, 2018b; Fossa, 2022) may get insights from Kant, who argues that autonomy implies "a free causality" (Kant, 1788: 5, 100) between actual rulers and actions—which makes rulers, i.e. humans, morally responsible for their actions and, consequently, morally "culpable and deserving of punishment" (Kant, 1788: 5, 100). Conversely, automation meant by "*automaton*" implies a "natural law of causality" (Kant, 1788: 5, 97) between causes (replacing actual rulers) and effects (replacing actions)—which deprives humans of being morally responsible for their actions and, consequently, morally "culpable and deserving of punishment". The loss of moral responsibility and its consequences may seem advantageous—more precisely, the advantageous shift of the dramatic burden of moral responsibility from humans to technology by making the former automated and the latter autonomous may be a promising way to read our technological era (as I have argued elsewhere[3] and shall continue to argue in the following chapters). Yet, even though depriving ourselves of the dramatic burdens of being morally responsible and "culpable and deserving of punishment" may ease our lives, Kant's insight is that we may lose what makes us actually having identities. In the case of the fatal crash, there is an essential difference between the following two scenarios. In the first scenario, we use "highly autonomous vehicles" and are not required to reflect upon the option to choose: trying to save our lives? Trying to save others' lives if we know them and care about them? Trying to save others' lives whoever they are? In the second scenario, we do not use "highly autonomous vehicles" and are required to reflect upon the option to choose—and reflecting upon the option to choose means shaping the identities that, finally, we can actually have. Conversely, the automatism between one cause and one effect means a randomness that cannot shape one's

[2] In the last years, increasing attention has been paid to a variety of ethical issues raised by autonomous vehicles, starting with the relationship between ethics and algorithms (whose complexity is stressed, for instance, by Goodall, 2014a, 2014b; Lin, 2015; Purves et al., 2015; Coeckelbergh, 2016; London & Danks, 2018; Santoni de Sio, 2017, but also Matthias, 2004, and whose complexity is challenged, for instance, through possible translations of ethical theories into algorithms by Powers, 2006, 2013; Gerdes & Thornton, 2015; Bonnefon et al., 2016; Coeckelbergh, 2016; Gurney, 2016; Gogoll & Müller, 2017; Leben, 2017; Awad et al., 2018; Faulhaber et al., 2019. When possible translations seem promising, ethical issues on mandatory ethics setting and nonmandatory ethics setting are raised especially by Millar, 2015; Contissa et al., 2017; Gogoll & Müller 2017; Faulhaber et al., 2019).

[3] Especially in Chiodo (2020a).

identity, which is made of one's analyses, abstractions, idealisations, values and, especially, accountabilities (which is the reason why, if you cannot be accountable, first, for your beliefs and, second, for your actions, then you "lose your integrity and so your identity [...]. That is, it is no longer to be able to think of yourself under the description under which you value yourself and find your life worth living and your actions worth undertaking").

The etymology of the word "automation" can make us go even further in distinguishing it from autonomy. Again, the words "automation" and "autonomy" share the first half of their etymology, i.e. αὐτός as "itself". But they do not share the second half of their etymology, which, in the case of "autonomy", is νόμος as "law", as we have seen, and, in the case of "automation", is (αὐτο)ματίζω (transliterated as *(auto)matizo*) as to "act of oneself, act offhand or unadvisedly", "to be done spontaneously or at random", "haphazard", to "introduce the agency of chance", "of things, [to] happen of themselves, casually", "to be self-produced" and, "of natural agencies, [to] act spontaneously".[4] In addition to the Greek verb (αὐτο)ματίζω, the Greek noun αὐτοματισμός (transliterated as *automatismos*) means "that which happens of itself, chance".[5] Moreover, the Greek noun αὐτόματον (transliterated as *automaton*), which is precisely the word used by Kant, means "accident".[6] Finally, the Greek noun Αὐτοματία (transliterated as *Automatia*) means "the goddess of chance",[7] who is defined by Smith as a "surname of Tyche or Fortuna, which seems to characterise her as the goddess who manages things according to her own will, without any regard to the merit of man"[8] (see also Murray, 1833: 577). If we move from the etymology of the word "automation" to its current definition, then its strict correlation with technology is totally explicit: "The action or process of introducing automatic equipment or device into a manufacturing or other process or facility; (also) the fact of making something (as a system, device, etc.) automatic. Originally (and

---

[4] Liddell-Scott-Jones Greek-English Lexicon, see http://stephanus.tlg.uci.edu/lsj/#eid=18225 (accessed in September 2021). It is worth noting that, as far as the contemporary ubiquitous use of the word "automation" is concerned, the sphere of human work's automation was key, specifically since 1947, when Ford Motor Company decided to "set up an 'automation department'. [...] [The] new unit hoped to increase the use of existing technologies – hydraulic, electromechanical, and pneumatic—to speed up operations and enhance productivity on the assembly line" (Rifkin, 1955: 66). As we shall see, both the etymological use and the philosophical use of the word "automation" are far more complex than its use in the sphere of human work. Yet, even the latter shows something essential in the former: the word "automation" emerges when a technological dimension replaces a human dimension not only for the purpose of improving the quality of human life but also, and especially, for the purpose of improving "speed", "operations", "productivity" and "the assembly line", whose relationship with the quality of human life may be dramatically dual.

[5] Liddell-Scott-Jones Greek-English Lexicon, see http://stephanus.tlg.uci.edu/lsj/#eid=18226 (accessed in September 2021).

[6] Liddell-Scott-Jones Greek-English Lexicon, see http://stephanus.tlg.uci.edu/lsj/#eid=18228 (accessed in September 2021).

[7] Liddell-Scott-Jones Greek-English Lexicon, see http://stephanus.tlg.uci.edu/lsj/#eid=18224 (accessed in September 2021).

[8] Smith 1867, see https://quod.lib.umich.edu/m/moa/acl3129.0001.001/462?page=root;rgn=full+text;size=100;view=image;q1=auto (accessed in September 2021).

now usually) in neutral sense, but in the 1950s often associated with the use of electronic or mechanical devices to replace human labour, and hence sometimes having negative connotations".[9] Thus, even though the words "automation" and "autonomy" share the first half of their etymology, they are definitely distinguished not only by the second half of their etymology but also by the current definition of autonomy. In the case of the word "autonomy", we find the capacity to "give oneself laws" that underpins one's identity and dignity, as we have seen. Conversely, in the case of the word "automation", on the one hand, we find that what is "self-given" is something "offhand", "unadvised", "spontaneous", "random", "haphazard", "introduc[ing] the agency of chance", "casual", "accident[al]" and "without any regard to the merit of man" and, on the other hand, we find something meant "to replace" something "human" (which is quite paradoxical from the following perspective: automation is meant "to replace" something "human", specifically "human labour", and ends up risking "replac[ing]" "human" autonomy itself, which underpins "human" identity and dignity, as we can continuously take note from the current way we use language).

Thus, the philosophical issue upon which we should reflect is the following: what does it mean to increasingly shift autonomy from humans to technologies and automation from technologies to humans if it means to increasingly replace, from humans' perspective, the capacity to "give oneself laws" with something "random"? More precisely, what does it mean to increasingly replace one's self-awareness as one's capacity to rule with randomness as merely causing an effect? Even more precisely, what does it mean to increasingly create technologies that merely cause effects coinciding with one's actions?

2.2 A Case in Point: Autonomous Vehicles

Autonomous vehicles are a case in point. Briefly, according to the SAE International's (formerly Society of Automotive Engineers') standard definition, autonomous vehicles are distinguished in six levels from zero to five (see at least Barabás et al., 2017). From zero to two, "You must constantly supervise these support features [i.e., progressively from zero to two, automatic emergency braking, blind spot warning, lane departure warning, lane centering, adaptive cruise control]; you must steer, brake or accelerate as needed to maintain safety".[10] From four to five, and partially in the case of three, "These automated driving features [i.e., progressively from three to five, traffic jam chauffeur, local driverless taxi, pedals] will not require you to take over driving".

According to the Waymo (formerly Google) Safety Report, "Waymo's mission is to bring autonomous driving technology to the world, making it safe and easy for

[9] Oxford English Dictionary, see https://www.oed.com/view/Entry/13468?redirectedFrom=automation#eid (accessed in September 2021).

[10] See, also for the following quotes, https://www.sae.org/news/2019/01/sae-updates-j3016-automated-driving-graphic (accessed in September 2021).

people and things to get where they are going. We're building the World's Most Experienced Driver and we believe our technology can improve access to mobility, and save thousands of lives now lost to traffic crashes" (Waymo Safety Report, 2021: 2). More precisely, "Our ultimate goal is to develop fully autonomous driving technology that can move people and things from A to B, anytime, anywhere, and in all conditions" (Waymo Safety Report, 2021: 16). It is worth noting that the words quoted above continuously stress an idea of shifting burdens from humans to autonomous vehicles: expressions of metaphorical lightness and absence of effort are ascribed to the former ("making it safe and easy for people" and "sav[ing] thousands of lives") and expressions of metaphorical heaviness and presence of effort are ascribed to the latter ("our technology can improve access to mobility, and save thousands of lives", and "fully autonomous driving technology [...] can move people and things from A to B, anytime, anywhere, and in all conditions"). Yet, something quite unexpected happens: "During our internal testing, however, we found that human drivers over-trusted the technology [level three] and were not monitoring the roadway carefully enough to be able to safely take control when needed. As driver-assist features become more advanced, drivers are often asked to transition from passenger to driver in a matter of seconds, often in challenging or complex situations with little context of the scene ahead. The more tasks the vehicle is responsible for, the more complicated and vulnerable this moment of transition becomes. Avoiding this 'handoff problem' is an important reason why Waymo is committed to fully autonomously driven vehicles. Our technology takes care of all of the driving, allowing passengers to stay passengers" (Waymo Safety Report, 2021: 13). At least two details are most interesting. First, humans' attitude, which may be described, again, in terms of shifting burdens from themselves to autonomous vehicles ("human drivers over-trusted the technology and were not monitoring the roadway carefully enough to be able to safely take control when needed"). Second, designers' attitude, which may be described in terms of even increasing the shift of burdens from humans to autonomous vehicles (if the issue is the "handoff problem", then the solution is "fully autonomously driven vehicles"). Thus, the relationship between humans and technology has the following balance: the latter is totally active ("technology takes care of all of the driving") and the former is totally passive (humans are "allow[ed] to stay passengers").[11] The current debate on autonomous vehicles starts questioning the solution proposed by reconsidering the balance of the relationship between humans and technology—we may say that what is a "problem" to solve from an engineering perspective may not be a "problem" to solve, but an opportunity to seize, from a philosophical perspective (for instance, not "to stay passengers" may be an opportunity to seize both in terms of epicurean driving pleasure and in terms of stoic exercising human capacities). According to the current debate, "fully autonomously driven vehicles", together with other autonomous technologies, may be at least suspicious, in that they "can pose a

[11] Interestingly enough, in the previous version of the Waymo Safety Report (2017: 13), the image of the section entitled *The Case for Full Autonomy. Allowing Passengers to Stay Passengers* meaningfully coincided with two human hands leaving the wheel, i.e. one of the symbols of human control (human hands: we even idiomatically say to have the situation in hand) leaving one of the symbols of control (the wheel: we even idiomatically say to be behind the wheel).

direct threat to a user's moral autonomy" (Millar, 2016: 790. See also Fossa, 2022) by "taking users entirely out of the ethical decision-making loop" (Millar, 2016: 790. See also Goodall, 2014a, 2014b; Himmelreich, 2018; Bellet et al., 2019; Cunneen et al., 2019; Cunneen et al., 2020). Thus, "we must step in and take the right actions to ensure that the human is the ultimate decision maker for autonomous systems" (Xu, 2021: 53): "Driverless does not, and should not, mean without a human operator" (Nunes et al., 2018: 171).[12] According to several authors, the most suspicious thing is a technological issue (for instance, the "handoff problem" when, typically in the case of level three, "drivers are often asked to transition from passenger to driver in a matter of seconds, often in challenging or complex situations with little context of the scene ahead", which is the reason why levels four and five may be safer than level three). But there is also a strictly philosophical issue to address, i.e. the meaning of the shift of burdens from humans to technology. As we have seen, the ancient Prometheus himself means that humans use technology to shift burdens from themselves to the latter. But our technological era is not only a matter of shifting concrete burdens from ourselves to technology (which happens, for instance, when the ancient Prometheus "brought brute beasts beneath the yoke to be subject to the collar and the pack-saddle, so that they might bear in men's stead their heaviest burdens") but also a matter of shifting abstract burdens from ourselves to technology (which happens, for instance, when the most authentic reason why we want "to stay passengers" is that we want to deprive ourselves of the dramatic burden of moral responsibility)—the more we move from the ancient Prometheus to our technological era, the more we shift from ourselves to technology not only our concrete burdens but also, and especially, our abstract burdens, especially the dramatic burden of moral responsibility, which we have to take whenever we have to make decisions.

As the greatest masterpieces in the history of Western literature show us, from Aeschylus himself to Sophocles to Shakespeare to Pirandello, the burden of moral responsibility, which we have to take whenever we have to make decisions, is actually most dramatic. Yet, it is also most essential to define, again, not only our identity but also our dignity: from Aeschylus himself to Sophocles to Shakespeare to Pirandello (see especially Aesch. *PB*, Soph. *OT*, Shakespeare, 1998; Pirandello, 1992 [1925]), humans, as they are thought of in Western culture, first, define their own identity especially according to their capacity to make the right decisions, from both a logical perspective and a moral perspective, even in the most difficult circumstances, from ancient culture characterised by τύχη (translated as *tyche*, which does not deprive Œdipus, for instance, of his dramatic burden of decision) to modern and contemporary culture characterised by free will (which makes the burden of decision even more dramatic), and, second, define their own dignity especially according to their capacity to bear the burden of the decisions they made. Our technological era seems to emerge as our greatest attempt to shift the burden of moral responsibility from ourselves to technology—our technological era seems to emerge as our greatest attempt to shift

[12] See also Fridman (2018), Calvert et al. (2019), Mecacci and Santoni de Sio (2019) and Stayton and Stilgoe (2020). I am grateful to Fabio Fossa for his hints on the current debate on autonomous vehicles.

what has been most essential for millennia to define our own identity and dignity from ourselves to technology.

The notion of autonomy is the greatest way in which modern and contemporary Western culture has addressed the issue of humans' identity and dignity as underpinned by their capacities, first, to make the right decisions and, second, to bear the burden of the decisions they make. If we move from Kant's philosophical work to contemporary philosophy, specifically to a most influential articulation of the notion of autonomy, then we can further read autonomous vehicles as a case in point of the shift of autonomy itself from ourselves to technology. According to the hierarchical notion of autonomy (see especially Frankfurt, 1971, 1999; Dworkin, 1988), we can have first-order volitions and second-order volitions: the former coincide with what we want and the latter coincide with what we want to want. For instance, on the one hand, we want to make our lives easier and, on the other hand, we want to want not to make our lives easier if it causes damage to someone. We may say that second-order volitions introduce a kind of idealisation, according to which we define ourselves not only in terms of what we really want but also, and especially, in terms of what we ideally want—and what we ideally want also means what kind of humans we want to be, i.e. what kind of identity we want to have, in that moving from wanting something to wanting to want something means moving from contingency to actually having a plan. Thus, we can actually plan our identities when we add second-order volitions to first-order volitions. Moreover, we can make our identities coincide with our actual plans when we have the capacity to make our first-order volitions coincide with our second-order volitions, i.e. to be the kind of humans we want to be, to have the kind of identity we want to have—which means to be autonomous. According to the hierarchical notion of autonomy, being autonomous can be thought of as the capacity to make second-order volitions underpin first-order volitions, i.e. to make the kind of identity we want to have (what we want to want) underpin the kind of identity we have (what we want). For instance, in the case of our example (on the one hand, we want to make our lives easier and, on the other hand, we want to want not to make our lives easier if it causes damage to someone), we are autonomous if we have the capacity not to make our lives easier if it causes damage to someone. If we go back to the Waymo Safety Report, then we should ask ourselves the following question: if "to stay passengers", i.e. to make our lives easier, would mean not to do everything possible to prevent a fatal crash (for instance, because of a lack of the software that cannot acknowledge an extremely rare circumstance), then would we want "to stay passengers", i.e. to make our lives easier? Our answer may be affirmative: we may autonomously decide not to do everything possible to prevent a fatal crash (which means that we may autonomously decide to expose ourselves to the following dramatic question that may happen to challenge ourselves in the future: did we make the right decision?). Conversely, we may answer that we autonomously decide to do everything possible to prevent a fatal crash, even though it may mean not to make our lives easier, i.e. not only not "to stay passengers" but also, and especially, not to shift from ourselves to technology the dramatic burdens of autonomously making decisions and acting in the most difficult circumstances. And the reason of our answer is precisely that we believe that the capacities to autonomously make decisions and act underpin who we want to be: humans whose ideal identity and dignity cannot do without the capacity to move from being metaphorically "passengers" to being metaphorically drivers. In both cases, we exercise a kind of autonomy: in the first case, the autonomy to actively decide "to stay passengers" and, in the second case, the autonomy to actively decide to be drivers.

But there is a third case, in which our autonomy, i.e. human autonomy, totally disappears: when we do not even ask ourselves the question on the relationship between what we want and what we want to want—and the more we trade our autonomy for technology's automation, the less we ask ourselves the question on the relationship between what we really want and what we ideally want, i.e. what kind of humans we want to be, what kind of identity we want to have. We may imagine "to stay passengers" in our "fully autonomously driven vehicles". We are likely to increasingly pay less attention to what happens between ourselves and the surrounding reality. The first time, we are likely to keep our eyes on the roads. Thus, when we happen to risk a collision, we can at least question ourselves: what is the right thing to do from our perspective? And what is the right thing to do from the pedestrians' perspective? Which means exercising a series of human capacities, from critical thinking to empathy (see Vallor, 2015, 2017). The second time, we are likely to partially take our eyes off the roads: we start reading, drinking and eating. The third time, we are likely to totally take our eyes off the roads: we start sleeping. Thus, when we happen to risk a collision, we cannot even question ourselves on what is the right thing to do from whoever's perspective. Which means increasingly atrophying a series of human capacities, from critical thinking to empathy. More precisely, the more we stop questioning ourselves, the more we trade our critical thinking and empathy for the software's output—but critical thinking and empathy are human capacities that require exercise not only to evolve but also not to atrophy. Finally, we end up acknowledging the software's output through a kind of act of faith (as we have seen in the case of algorithms' inscrutability)—and acknowledging whoever's output through a kind of act of faith means not being autonomous (from both Kant's perspective, according to which we are not autonomous at all if what we do is not underpinned by our own reason, and the perspective of the hierarchical notion of autonomy, according to which we are not autonomous at all if what we do is not underpinned by what we want to do). Even more precisely, we may say that the more we stop questioning ourselves, the more we trade our autonomy for the software's automation—the more we use technology to stop questioning ourselves, the more we shift autonomy from ourselves to technology and automation from technology to ourselves.

The overturning described above is something we started when, mythologically speaking, the ancient Prometheus gave us Hephaestus' fire and Athena's wisdom in the arts. I myself use technology not only to trade my capacities in general (for instance, my memory) for technology's automation (for instance, for my agenda) but also to trade my autonomy in particular for technology's automation, which happens, for instance, whenever I google something, in that I frequently rely on what Google's algorithm makes appear on the first page. Yet, I realise that the more I deprive myself of circumstances in which I exercise my capacities in general and my autonomy in particular, the more I risk atrophying them—also considering that, since autonomy implies a most sophisticated exercise of reason, as Kant masterfully teaches us, we should take great care of its exercise.

Thus, where should we fix the limit between what we should shift and what we should not shift? There are at least two kinds of answers. On the one hand, we should answer as single individuals—which means that we, as single individuals, should decide what human capacities are at the essence of who we, as single individuals, want to be. For instance, I may decide that at the essence of my identity there should be my capacities, first, to make decisions and, second, to bear the burden of the decisions I make, but not my memory. On the other hand, we should answer as a society—which means that we should decide not only what human capacities are at the essence of the kind of society we want to shape but also what policies can make us shape it. For instance, we may decide that at the essence of our society there should be our capacity to be altruistic, even as "passengers", and not egoistical and impartial (see Contissa et al., 2017).

2.3 Autonomous Vehicles and the Moral Machine

The reason why philosophers should team up with engineers more than ever is precisely the following: our current relationship with technology raises crucial philosophical questions to our future as humans, from our identity as single individuals to our identity as a society. Autonomous vehicles offer at least another example to reflect upon crucial philosophical questions on our present and future identity: the moral machine, designed by the Massachusetts Institute of Technology as a digital platform asking moral dilemmas raised by autonomous vehicles (and answered by millions of people from more than two hundred countries). The moral machine proposes a variety of scenarios and asks "moral dilemmas, where a driverless car must choose the lesser of two evils, such as killing two passengers or five pedestrians. As an outside observer, you judge which outcome you think is more acceptable. You can then see how your responses compare with those of other people"[13] (see especially Bonnefon et al., 2016; Shariff et al., 2017; Awad et al., 2018, 2020a, 2020b). The moral machine proposes scenarios raising not only questions on the relationship between passengers and pedestrians (implying reflections upon what is fair if we take the perspective of "an outside observer", i.e. if we do not identify ourselves with passengers, on the one hand, and pedestrians, on the other hand) but also questions on a greater variety of moral dilemmas (implying reflections upon what is fair if we consider diversities characterising societies). More precisely, moral dilemmas focus on "nine factors: sparing humans (versus pets), staying on course (versus swerving), sparing passengers (versus pedestrians), sparing more lives (versus fewer lives), sparing men (versus women), sparing the young (versus the elderly), sparing pedestrians who cross legally (versus jaywalking), sparing the fit (versus the less

[13] See https://www.moralmachine.net/ (accessed in October 2021).

fit), and sparing those with higher social status (versus lower social status)" (Awad et al., 2018: 60). People's answers were analysed and three "moral clusters" (Awad et al., 2018: 61) were identified: Western "moral cluster" (in which preferences for inaction are higher and preferences for sparing women are lower), Eastern "moral cluster" (in which preferences for sparing pedestrians are higher and preferences for sparing the young are lower) and Southern "moral cluster" (in which preferences for sparing people with higher social status are higher and preferences for inaction are lower). If we analyse people's answers from a global perspective, then three highest preferences are identified: "sparing humans over animals, sparing more lives, and sparing young lives" (Awad et al., 2018: 60), which makes the authors of the analysis say that "these three preferences may be considered essential building blocks for machine ethics, or at least essential topics to be considered by policymakers" (Awad et al., 2018: 61), in that they "can trace our path to universal machine ethics" (Awad et al., 2018: 61. On machine ethics in general see at least Moor, 2006; Wallach & Allen, 2008; Anderson & Anderson, 2011; Goodall, 2014b). Yet, I think that a crucial philosophical question arises: if we reduce our actions' rules to a merely particular dimension, i.e. as random as our contingent answers, then what is the destiny of our autonomy?

The example of the moral machine is most interesting as a clear symptom of a more general phenomenon that has to do, again, with the immanentisation of the ruler: if it is true that technology should make our lives easier by depriving ourselves of burdens, it is also true that, if rules are needed, then, first, they should be easy to find and, second, they should be easily immanentised by technology itself—the example of the moral machine is a clear symptom of our use of technology to deprive ourselves of burdens by immanentising the ruler.

Taking the perspective of "an outside observer" is not sufficient to escape from a merely particular dimension, in that it does not make us escape from our answers' contingent, and even idiosyncratic, randomness (for instance, we may happen to answer that we spare passengers versus pedestrians because we have recently had an accident as passengers on account of a pedestrian. And we may happen to answer that we spare pedestrians versus passengers because we have recently had an accident as pedestrians on account of a passenger). Conversely, escaping from a merely particular dimension means, for instance, what follows: "Act as if the maxim of your action were to become by your will a universal law of nature" (Kant, 1785: 4: 421). And, to refer to a philosopher closer to our technological era, we may say that escaping from a merely particular dimension means, for instance, what follows (which is not as powerful as what Kant argues): our preferences should be underpinned by "a veil of ignorance" (Rawls, 1999 [1971]: 11), i.e. by imagining "that no one knows his place in society, his class position or social status, nor does anyone know his fortune in the distribution of natural assets and abilities, his intelligence, strength, and the like. I shall even assume that the parties do not know their conceptions of the good or their special psychological propensities. The principles of justice are chosen behind a veil of ignorance" (Rawls, 1999 [1971]: 11). If we reflect upon the destiny of our autonomy in the two kinds of scenarios proposed above, then we should say that, as far as the scenario of "moral clusters" is concerned, which globally result in "sparing humans over animals, sparing more lives, and sparing young lives", we should ask ourselves the following question: would we be ready, for instance, to think of "sparing the young (versus the elderly)" as a "universal machine ethics"? I think that several of us would not be perfectly ready. More precisely, I think that several of us would feel a kind of violation of freedom: what if we ourselves are the elderly? What if our loved ones are the elderly? If we move from the scenario of "moral clusters" to the scenario of a kind of idealisation, which is proposed by both Kant and Rawls, then something crucial changes: several of us are likely to agree that, if we use both the former's categorical imperative and the latter's "veil of ignorance", then the right thing to do is to spare the lawful (in the moral machine's terms, to spare, for instance, "pedestrians who cross legally (versus jaywalking)"). If it is true that several of us are likely to be perfectly ready to agree on it and, more precisely, if it is true that several of us are likely not to feel a kind of violation of freedom, then what is the reason of the difference between the two kinds of scenarios proposed above? I believe that we may explain the difference in terms of autonomy, which is sabotaged in the former case and enhanced in the latter case.

In the case of the first scenario, a contingent, and even idiosyncratic, randomness, which coincides with nothing but an empirical sum, becomes a universal rule. But an empirical sum cannot become a universal rule, in that it is likely to exclude minorities, which would necessarily be violated[14]: again, what if we ourselves are the elderly? What if our loved ones are the elderly? The key issue is that, if we ourselves are the elderly, then we cannot change our condition at all, and, if our loved ones are the elderly, then they cannot change their condition at all. Moreover, an empirical sum that becomes a rule is likely to work as something heteronomously imposed precisely because, being an empirical content coinciding with a condition existing outside our possibility of choosing, it cannot be our autonomy's domain.

Conversely, in the case of the second scenario, a kind of idealisation becomes a universal rule. Speaking of a kind of idealisation means speaking of what follows: if we ask ourselves what is "the maxim of [...] [our] action" that can "become by [...] [our] will a universal law of nature", then we can answer that "the maxim" according to which the right thing to do is to spare the lawful, whoever they are, can "become

[14] It is worth noting that, at least according to the European recommendations, empirical sums seem excluded as follows: "CAVs [connected and automated vehicles] should be designed and operated in ways that neither discriminate against individuals or groups of users, nor create or reinforce large-scale social inequalities among users. They should also be designed in a way that takes proactive measures for promoting inclusivity" (Horizon 2020 Commission Expert Group, 2020: 8). More precisely, "If such discriminatory user segmentation and differential service provision is connected to individuals or groups of users on grounds directly or indirectly related to, for example, race, gender, social class, income, religion, sexual orientation, place of residence, citizenship, political conviction or religious belief, there will be legal, as well as ethical, violations of principles of non-discrimination and fairness" (Horizon 2020 Commission Expert Group, 2020: 44). The German recommendations, which anticipate the European recommendations, also seem to exclude empirical sums as follows: "In the event of unavoidable accident situations, any distinction based on personal features (age, gender, physical or mental constitution) is strictly prohibited. It is also prohibited to offset victims against one another" (Federal Minister of Transport and Digital Infrastructure, 2017: 7).

by […] [our] will a universal law of nature". The key issue is that, whoever we are, we can change our condition, for instance, by moving from jaywalking to being pedestrians who cross legally—whoever we are, we can autonomously change our actions without heteronomous impositions precisely because we give ourselves a rule that, being a form that coincides with no condition existing outside our possibility of choosing, can be our autonomy's domain: I myself happen to jaywalk, but I can autonomously give myself the rule according to which the right thing to do is to spare the lawful precisely because "the maxim of […] [my] action [that] were to become by […] [my] will a universal law of nature", being formal, and not empirical, can be not only my autonomy's domain, without violating my freedom, but also whoever's autonomy's domain (which also protects me from others' jaywalking, in that both I and others have excellent reasons to stop jaywalking: reciprocity of rules).

The reason why the increasing technological immanentisation of rules (which strictly correlates with epistemological and ethical anarchism) is dangerous is precisely the following: the more we make rules lose universality, i.e. a kind of idealisation that makes them formal, the more we risk falling into heteronomous rules that work against us, i.e. our autonomy and freedom, in that rules coinciding with particularity can be nothing but contingent, and even idiosyncratic—the increasing technological immanentisation of rules make us risk falling into the randomness that results from nothing but empirical sums, which technology's ease of computing surely strengthen, as we have seen in the case of the moral machine.

More generally, we should be aware that technology's ease of computing, together with the divine prerogatives (omnipresence, omniscience, omnipotence and inscrutability) it increasingly obtains, equally increases a severe risk: taking mere statistics, i.e. kinds of empirical sums, instead of rules. Yet, again, rules resulting from merely particular dimensions can give us nothing but the randomness of contingency, and even of idiosyncrasy—moreover, statistics taken instead of rules can do nothing but work as something heteronomously imposed on our autonomy and freedom: there is no self-given rule at all if I take instead of it a statistic according to which, for instance, I must take medicine tested on males, on whom the statistic is based, but I am a female (which happens, see at least Dusenbery, 2018).

Objections to my argument may be raised. The first objection may be the following: what happens when both parties are the lawful to spare? My answer is that we should accept the complexity of reality: sometimes one party is right and one party is wrong, which makes things easier, and sometimes both parties are right, which makes things more difficult. But we should not accept that, if things are more difficult, then we should stop exercising our autonomy and start relying on technology's automation by making the latter replace the former whenever we are afraid of bearing the burden of moral responsibility if we turn out to be wrong—exercising our autonomy makes sense in itself if we think of it as a human capacity that, more than others, is at the essence of who we want to be. If we write autonomous vehicles' algorithms, then it is not always easy to make them spare the lawful versus jaywalking, in that we know that it also means sparing a killer who crosses legally versus a child who crosses illegally. Yet, exercising our autonomy is one of the greatest things we can do as humans, in that it means that, also in the most difficult circumstances, we do not let things fall into the unaccountability of randomness, specifically technology's unaccountability (see at least Mittelstadt et al., 2016; Coeckelbergh, 2020b. See also Hevelke & Nida-Rumelin, 2015; Liu, 2017; Dietrich & Weisswange, 2019; Epting, 2019)—conversely, we have both the intellectual courage and the moral courage to be personally accountable both in front of ourselves and in front of others, whom we do not leave in the dramatic anonymity of technology's unaccountability that cannot offer listening, understanding, answers and solutions. The second objection may be the following: since there are authors working on writing algorithms that embed ethical theories, technology, specifically autonomous vehicles' algorithms, not only does not sabotage autonomy but also enhance it (see, for instance, Powers, 2006 on Kant and Leben, 2017 on Rawls). My answer is that we, as users of autonomous vehicles, would "stay passengers": if it is true that algorithms embedding promising ethical theories can be equally promising, at least more promising than algorithms embedding empirical sums, it is also true that no one but the professionals who write them would not metaphorically "stay passengers", i.e. would literally exercise their autonomy. It is no coincidence that one of the issues characterising the current debate on autonomous vehicles is the distribution of responsibility, which starts being thought of as something that has to do not only with the professionals who write algorithms and, more generally, develop autonomous vehicles but also with users, and even governments (see at least Nissenbaum, 1996; van de Poel et al., 2015). According to the European recommendations on CAV, i.e. connected and automated vehicles, we should "Ensure accountability for the behaviour of CAVs (duty to explain). 'Accountability' is here defined as a specific form of responsibility arising from the obligation to explain something that has happened and one's role in that happening. A fair system of accountability requires that: (a) formal and informal fora and mechanisms of accountability are created with respect to CAVs; (b) different actors are sufficiently aware of and able to discharge their duty to justify the operation of the system to the relevant fora; (c) and the system of which CAVs are a part is not too complex, opaque, or unpredictable" (Horizon 2020 Commission Expert Group, 2020: 10). Thus, we should "Promote a fair system for the attribution of moral and legal culpability for the behaviour of CAVs. The development of fair criteria for culpability attribution is key to reasonable moral and social practices of blame and punishment—e.g. social pressure or public shaming on the agents responsible for avoidable collisions involving CAVs—as well as fair and effective mechanisms of attribution of legal liability for crashes involving CAVs" (Horizon 2020 Commission Expert Group, 2020: 10). As far as the reference to autonomy in particular is concerned, "Human beings should be seen as free moral agents, who demand respect for the conditions of their agency. This requires that CAVs should protect and promote human beings' capacity to decide about their movements and, more generally, to set their own standards and ends for accommodating a variety of conceptions of a 'good life'" (Horizon 2020 Commission Expert Group, 2020: 22). As far as the reference to responsibility in particular is concerned, "The counterpart of human autonomy is human responsibility. Both individual persons and institutional stakeholders can and should be held morally and legally responsible for the consequences of their actions when it is appropriate to do so. At the same time, they should be given a fair capacity and opportunity to behave according to moral and legal expectations. In relation to CAVs, this means establishing clear moral and legal standards of responsibility, while at the same time providing different actors (CAV users, but also CAV manufacturers and deployers) with sufficient knowledge, capacity, motivation and opportunities to comply with these standards" (Horizon 2020 Commission Expert Group, 2020: 22).[15] Both the efforts against "too complex, opaque, or unpredictable" systems and the efforts for accountability, autonomy and responsibility are noteworthy. Yet, even though autonomous vehicles may give us great opportunities, first, to reduce accidents, second, to make mobility more inclusive (by making, for instance, disabled people move in an easier way) and, third, to make mobility more sustainable (by both decreasing the number of vehicles and increasing energy efficiency), philosophical questions may arise (see at least Bradshaw-Martin & Easton, 2014; Taiebat et al., 2018; Martin, 2020; Banerjee, 2021).

First, as far as accountability in general is concerned, even though the effort to guarantee it is surely remarkable, a philosophical question may arise on the difference between accountability as something distributed between several actors (developers, users and governments) and accountability as something attributed to a single individual. It is no coincidence that the dramatic burden of moral responsibility, which we have to take whenever we have to make decisions, has been represented, from Aeschylus to Sophocles to Shakespeare to Pirandello, as a single individual's burden, and not as a distributed burden: the tragic heroes of Western culture are, for instance, Œdipus and Hamlet, and not the Greek society and the Danish society. On the one hand, Western culture has stressed for millennia the role of the single individual versus the role of whatever society. On the other hand, if the burden does not vary, but its distribution varies, then the burden is heavier when it is attributed to a single individual, i.e. borne by a single individual, and lighter when it is distributed between several actors, i.e. borne by whatever society—at least from a typically Western perspective, the test humans have to pass to define their own identity and dignity is Œdipus' and Hamlet's, i.e. to train the capacity to bear the heaviest burden, which is

[15] The German recommendations also refer to autonomy and responsibility. As far as the former is concerned, "Technological development obeys the principle of personal autonomy, which means that individuals enjoy freedom of action for which they themselves are responsible" (Federal Minister of Transport and Digital Infrastructure, 2017: 6). As far as the latter is concerned, "The personal responsibility of individuals for taking decisions is an expression of a society centred on individual human beings, with their entitlement to personal development and their need for protection. The purpose of all governmental and political regulatory decisions is thus to promote the free development and the protection of individuals. In a free society, the way in which technology is statutorily fleshed out is such that a balance is struck between maximum personal freedom of choice in a general regime of development and the freedom of others and their safety" (Federal Minister of Transport and Digital Infrastructure, 2017: 6). Finally, as far as the issue of accountability is concerned, "the accountability that was previously the sole preserve of the individual shifts from the motorist to the manufacturers and operators of the technological systems and to be the bodies responsible for taking infrastructure, policy and legal decisions. Statutory liability regimes and their fleshing out in the everyday decisions taken by the court must sufficiently reflect this transition" (Federal Minister of Transport and Digital Infrastructure, 2017: 7) and "The distribution of responsibilities (and thus of accountability) [...] should be documented and stored" (Federal Minister of Transport and Digital Infrastructure, 2017: 9).

attributed to one actor, and not distributed between several actors: the burden of individual accountability. The European recommendations' effort to attribute "moral and legal culpability for the behaviour of CAVs" is remarkable. Yet, at least from a philosophical perspective, "moral and social practices of blame and punishment—e.g. social pressure or public shaming on the agents responsible for avoidable collisions involving CAVs—as well as fair and effective mechanisms of attribution of legal liability for crashes involving CAVs", are more easily bearable when they are not only distributed but also shared. If I, as a user of an autonomous vehicle, am culpable for a fatal crash together with its developers and our government, then the burden I have to bear is not only lighter from my own perspective, in that it is distributed, but also less dramatic from my own perspective, in that it is shared—and the role played by my own perspective is crucial to define my own identity and dignity. Even proverbially, a trouble shared is a trouble halved: being culpable for a fatal crash together with, for instance, the biggest global company of autonomous vehicles and a European government agreeing on the European recommendations can make my own accountability, autonomy and responsibility equally halved. If even the biggest global company of autonomous vehicles and my government agreeing on the European recommendations make mistakes, then why should I think that I can actually make no mistakes? Thus, the distribution of accountability is dual: it is both a remarkable effort not to make accountability disappear and at least a potential risk of ending up decreasing individual accountability and responsibility—which also means decreasing individual autonomy as its "counterpart" ("The counterpart of human autonomy is human responsibility").

Second, as far as autonomy in particular is concerned, a philosophical question may arise on the relationship between what is real and what is ideal. As far as the latter is concerned, "Human beings should be seen as free moral agents, who demand respect for the conditions of their agency. This requires that CAVs should protect and promote human beings' capacity to decide about their movements and, more generally, to set their own standards and ends". From an ideal perspective, I think that, again, what is written is remarkable. Yet, from a real perspective, I think that it is more difficult for me "to set [...] [my] own standards and ends" as a passenger than as a driver, in that I already know that, at least instinctively, whenever technology can make me move from being metaphorically a driver to being metaphorically a passenger, the latter role wins over the former role for at least the following reasons: first, it is easier and quicker (and I am in a hurry) and, second, knowing that, again, I share a decision with several actors, who are implied by technology itself (from developers to governments to other users) is more reassuring than knowing that I decide totally for myself. And, in addition to possible psychological reasons, it happens to me for both epistemological reasons (sometimes I happen to think that a decision judged as fair by several actors is likely to be fairer than a decision judged as fair by me) and practical reasons (again, sometimes I happen to think that distributing accountability between me and several actors is more bearable than attributing accountability to me). Thus, the more I shift burdens from myself to technology, the more "to set [...] [my] own standards and ends" is difficult for me, in that what really happens to me is that the more I shift burdens from myself to technology, the more I automate whatever I do—and the more I automate whatever I do, the less I exercise my autonomy. If the fatal crash results from my decision and action, then I am likely to challenge my autonomy for the rest of my life. Conversely, if the fatal crash results from technology's decision and action, i.e. decision and action shared by several actors and automated, then I am likely to halve, at least, the burden of my challenge with my autonomy.

Third, as far as responsibility in particular is concerned, a philosophical question may arise on the meaning of "when it is appropriate to do so" in the following words: "Both individual persons and institutional stakeholders can and should be held morally and legally responsible for the consequences of their actions when it is appropriate to do so". Moreover, the philosophical question on duality raised by the case of accountability also arises in the case of responsibility: the more responsibility is distributed (which is in itself a remarkable effort not to make responsibility disappear), the more we risk, at least potentially, ending up decreasing individual responsibility. Yet, as the greatest masterpieces in the history of Western literature show us, it is precisely the burden of individual responsibility what can make us evolve more radically, in that we do not experience a kind of reassuring sharing, but a kind of dramatically challenging sense of responsibility, in the best case scenario, and a kind of dramatically challenging sense of guilt, in the worst case scenario.

Finally, a further philosophical question may arise: more generally, by increasingly relying on technological solutions whenever we have to make decisions, what if we increasingly stop questioning ourselves on what is the right thing to do? In the case of autonomous vehicles in particular, if it is true that we are likely to pay close attention to the contract to sign when we buy our autonomous vehicle, it is also true that, by increasingly relying on autonomous driving, we are likely to pay close attention to something else, from reading to drinking to eating to sleeping—we are likely to merely rely on our autonomous vehicles, which means that, again, we trade our autonomy for technology's automation. But what if we are drivers? As drivers, we are likely to always pay close attention to what happens between ourselves and the surrounding reality: the exception happening to us as single individuals when we drive definitely wins over the standard shared by us as a social network of developers, users and governments—as drivers, we can do nothing but exercise our conscious decision and action, i.e. autonomy, in that the price to pay for our individual mistakes is exceedingly high not only epistemologically and ethically but also practically and legally (yet, in return, our autonomy, by being increasingly exercised, improves). In the case of technology in general, the following chapter shall also offer further examples to reflect upon the philosophical question on the destiny of our autonomy if, by increasingly relying on technological solutions whenever we have to make decisions, we increasingly stop questioning ourselves on what is the right thing to do. But, before proceeding with the following chapter, we should add one last key question: what may be the reason why, by increasingly relying on technological solutions whenever we have to make decisions, we trade our autonomy for technology's automation?

2.4 Trading Humans' Autonomy for Technology's Automation as Scapegoating

Relatively recently, i.e. without going back to Kant's philosophical work on the notion of autonomy, Berlin wrote: "I wish my life and decisions to depend on myself, not on external forces of whatever kind. I wish to be the instrument of my own, not of other men's, acts of will. I wish to be a subject, not an object; to be moved by reasons, by conscious purposes, which are my own, not by causes which affect me, as it were, from outside. I wish to be somebody, not nobody; a doer—deciding, not being decided for, self-directed and not acted upon by external nature or by other men as if I were a thing, or an animal, or a slave incapable of playing a human role. That is, of conceiving goals and policies of my own and realising them. This is at least part of what I mean when I say that I am rational, and that it is my reason that distinguishes me as a human being from the rest of the world. I wish, above all, to be conscious of myself as a thinking, willing, active being, bearing responsibility for my choices and able to explain them by reference to my own ideas and purposes. I feel free to the degree that I believe this to be true, and enslaved to the degree that I am made to realise that it is not" (Berlin, 1997: 203). We may reflect upon Berlin's inspiring words for days. And we may start with the following question: what happened to autonomy, celebrated by moving words in the second half of the last century and increasingly shifted from humans to technology in the last years?

A working hypothesis may be that autonomy, after its own irresistible rise especially from modern culture (from Kant onwards, as we have seen) to contemporary culture, started becoming its own worst enemy: what burden could be heavier than the burden of an exceedingly celebrated autonomy? More precisely, if autonomy even defines human identity and dignity, then what burden could be heavier than autonomy itself for humans? Being "a doer" is wonderful and terrifying at the same time. It is wonderful "to be moved by reasons, by conscious purposes, which are my own", but what if my "reasons" and "purposes" are dramatically wrong? It is wonderful "to be conscious of myself as a thinking, willing, active being, bearing responsibility for my choices and able to explain them by reference to my own ideas and purposes", but what if "my choices", "ideas and purposes" are dramatically wrong? Would it not be less terrifying "to be moved by" "causes which affect me, as it were, from outside"? Would it not be less terrifying to be effects of causes, and not causes of effects? A working hypothesis may be that one of the key reasons why we increasingly trade our autonomy for technology's automation is precisely the following: the more our autonomy is on the rise, the heavier, and even unbearable, the burden we have to bear both epistemologically and ethically is—paradoxically enough, the more our autonomy is on the rise, the more we trade it for what can make us move from being causes of effects to being effects of causes. Is technology, specifically technology's automation, our perfect solution?

Interestingly enough, in years characterised by autonomy's irresistible rise in contemporary philosophy, Reichenbach rewrites Shakespeare's figure of Hamlet as follows: "To be or not to be—that is not a question but a tautology. I am not interested in empty statements. I want to know the truth of a synthetic statement: I want to know whether I shall be. Which means whether I shall have the courage to avenge my father? Why do I need courage? It is true, my mother's husband, the king, is a powerful man and I shall risk my life. But if I can make it plain to everybody that he murdered my father, everybody will be on my side. If I can make it plain to everybody. It is so plain to me. Why is it plain? I have good evidence. The ghost was very conclusive in his arguments. But he is only a ghost. Does he exist? I could not very well ask him. Maybe I dreamed him. But there is other evidence. That man had a motive to kill my father. What a chance to become king of Denmark! And the hurry with which my mother married him. My father had always been a healthy man. It's a good piece of indirect evidence. But that's it: nothing but indirect evidence. Am I allowed to believe what is only probable? Here is the point where I lack the courage. It is not that I am afraid of the present king. I am afraid of doing something on the basis of a mere probability. The logician tells me that a probability has no meaning for an individual case. How then can I act in this case? That is what happens when you ask the logician. The native hue of resolution is sicklied over with the pale cast of thought. But what if I should start thinking after the deed and find out I should not have done it? Is the logician so bad? He tells me that if something is probable I am allowed to make a posit and act as though it were true. In doing so I shall be right in the greater number of cases. But shall I be right in this case? No answer. The logician says: act. You will be right in the greater number of cases. […] I shall know it for certain? There is no certainty. The probability will be increased and my posit will have a higher rating. I can count on a greater percentage of correct results. That is all I can reach. I can't get away from making a posit. I want certainty, but all the logician has for me is the advice to make posits. There I am, the eternal Hamlet. What does it help me to ask the logician, if all he tells me is to make posits? His advice confirms my doubt rather than giving me the courage I need for my action" (Reichenbach, 1951: 250–251). The reason why I extensively quote Reichenbach's figure of Hamlet is that I believe that it can perfectly represent one of the key reasons why our autonomy quickly moves from being on the rise to being in crisis. We ourselves are "the eternal Hamlet", in that our era's complexity and uncertainty make our autonomy's burden heavier, and even unbearable, both epistemologically and ethically. From globalisation's complexity to pandemic's uncertainty, we ourselves are "the eternal Hamlet" who has to face, from an epistemological perspective, the lack of certainty, in that "There is no certainty": "good evidence" and "arguments" give me nothing but "probability", which means that "my posit will have a higher rating" and "I can count on a greater percentage of correct results"—but the Hamletic question is "Am I allowed to believe what is only probable?". Consequently, from an ethical perspective, we ourselves are "the eternal Hamlet" who has to face the lack of "the courage I need for my action", in that "I am afraid of doing something on the basis of a mere probability": even though the logician "tells me that if something is probable I am allowed to make a posit and act as though it were true", in that "In doing so I shall be right in the greater number of cases", he also "tells me that a probability has no meaning for an individual case"—thus, the Hamletic questions are "But shall I be right in this case?", "How then can

I act in this case?" and "what if I should start thinking after the deed and find out I should not have done it?". But there is "No answer".

Is technology, specifically technology's automation, our perfect "answer"? From autonomous vehicles to, for instance, "predictive algorithm[s]" autonomously deciding on our health to algorithms autonomously deciding on our universities' ranking, we seem to increasingly answer the Hamletic questions as follows: if we lack epistemological certainty and ethical courage, then we can make technology decide for us, in that, on the one hand, its way to decide, i.e. computation, has an excellent reputation in our culture (see at least Porter, 1995) and, on the other hand, if something goes wrong ("what if I should start thinking after the deed and find out I should not have done it?"), then our burden is exceedingly lighter, from being distributed between several actors to being totally dissolved into technology ("if I should start thinking after the deed and find out I should not have done it", then my "answer" can be that algorithms decided)—if something goes wrong, then technology can be our perfect scapegoat to which we can shift our epistemological and ethical burdens, from the most dramatic ones to the easiest ones.

Yet, we know that algorithms are likely to share Hamlet's lack: even though they process more data more quickly than Hamlet, the kind of knowledge they can give us is nothing but "a mere probability", specifically "a higher rating" and "a greater percentage of correct results"—algorithms are likely to share Hamlet's lack, in that they can give us "no certainty": again, if "a probability has no meaning for an individual case", then how "can I act in this case?". Thus, the reason why we continue to replace our decisions with algorithms' decisions cannot totally coincide with the kind of knowledge they can give us: the kind of knowledge does not change, in that, in both cases, it is "a mere probability", which means that, even though it can have "a higher rating" and "a greater percentage of correct results" in the case of algorithms, it can have "no meaning for an individual case"—and whatever case we face, it can be nothing but "an individual case", of which algorithms can give us "no certainty".

Thus, why do we continue to replace our decisions with algorithms' decisions, if the kind of knowledge they can give us is not essentially different from the kind of knowledge we ourselves can give ourselves? I believe that the rise of our use of the word "autonomy" to define technologies in the last years can be a meaningful clue (as we have seen, autonomous systems, autonomous software, autonomous devices, autonomous applications, autonomous machines, autonomous vehicles, autonomous equipment, autonomous drones, autonomous weapons, autonomous robots, autonomous agents, autonomous workloads and so forth), which leads us to think that there is something more essential than algorithms' quicker process of more data (otherwise, we would not risk ending up acknowledging their outputs through a kind of act of faith, as we have seen). More precisely, if we combine the two irresistible rises upon which we have reflected, i.e. the rise of autonomy especially from modern culture to contemporary culture and the rise of our use of the word "autonomy" to define technologies in the last years, then the paradox that emerges is quite clear: we give technologies nothing less than the most successful definition of humans (at least in modern and contemporary Western culture)—and shifting from ourselves to technologies nothing less than our most successful definition as humans means that a kind of paradoxical externalisation of our own core identity emerges.

But why could we paradoxically externalise our own core identity by depriving ourselves, for instance, of Berlin's touching definition, according to which "I wish, above all, to be conscious of myself as a thinking, willing, active being, bearing responsibility for my choices and able to explain them by reference to my own ideas and purposes"? My working hypothesis, partially explored and partially to explore in the following chapters, is that the irresistible rise itself of autonomy as humans' core identity becomes an unbearable burden: "bearing responsibility for my choices and [being] able to explain them by reference to my own ideas and purposes" becomes an unbearable burden if, as Reichenbach equally touchingly writes, "There is no certainty" and, consequently, I lack "the courage I need for my action"—the irresistible rise itself of autonomy as humans' core identity becomes a kind of epistemological and ethical paralysis. Thus, why do we not create something that can give us not only "a higher rating" and "a greater percentage of correct results" but also, and especially, a perfect escape from autonomy's unbearable epistemological and ethical burdens, especially if something goes wrong?

Autonomy can be thought of as something we potentially have as humans, but not as something we actually have as humans. From Kant to Berlin, being actually autonomous requires effort. To apply the former's categorical imperative, we should exercise, for instance, the capacity to abstract, i.e. to move from what is particular to what is universal. And to apply the latter's words, we should exercise, for instance, the capacity "to explain" our "choices" by referring to our "ideas and purposes". Conversely, when our autonomous weapons kill for us, we may unburden ourselves of our effort, i.e. our epistemological and ethical burdens. Even though several authors stress the need not to sabotage human control (see at least Roff & Moyes, 2016; Santoni de Sio & van den Hoven, 2018; Amoroso & Tamburrini, 2021), especially through the notion of meaningful human control, philosophical questions may arise. First, what are the key reasons why we shift killing in a war from ourselves to technology? Are more restriction, more precision and sparing selected categories of humans satisfactory reasons? Or should we think that there is something more, specifically the attempt to unburden ourselves of our unbearable burdens, in the definition itself of autonomous weapons as "A weapon system that, once activated, can select and engage targets without further intervention by a human operator" (United States Department of Defense, 2012: 13)? Second, what are the key reasons why, even in cases in which human control is thought of as needed, it ends up being distributed between several actors, which may also mean that it risks dissolving into technological bureaucracy's inscrutability? Or should we think that technological bureaucracy's inscrutability is also, and especially, our perfect way to escape, again, from our unbearable burdens?

Autonomy as humans' core identity is wonderful and terrifying at the same time, in that it exposes humans to both their best individual ups (for instance, when they are considered as the best possible individuals for having consciously saved someone) and their worst individual downs (for instance, when they are considered as the worst possible individuals for having consciously killed someone). Thus, shifting autonomy from humans to technology may mean a kind of strengthened scapegoating. On the one hand, when things go wrong, humans can say that it is technology's fault (for instance, in the case of an algorithm's output according to which someone is not eligible to obtain something). On the other hand, when things go wrong, humans can say that it is no one's fault (for instance, in the case of an output whose inputs, distributed between several actors, dissolve into technological bureaucracy's inscrutability). In both cases, humans unburden themselves of their unbearable burdens, which are not only epistemological and ethical burdens but also the burden of their own individuality as a further essential cornerstone of Western culture—shifting autonomy from humans to technology especially means shifting individual autonomy from humans to technology: technology as humans' perfect scapegoat especially means that fault can more easily be technology's, no one's and, in the worst case scenario, common, but it cannot more easily be individual's.

Chapter 3
Second Case in Point: Self-Tracking Technologies and the Quantified Self

3.1 The Destiny of the Self in Technologies of the Self

The end of the last chapter was on the clue according to which technology can be thought of as our perfect scapegoat, in that the more we trade our autonomy for technology's automation, the less we bear not only epistemological and ethical burdens but also the burden of being single individuals—the burden of being dramatically alone when it comes to facing our own faults, in that, through technology as our perfect scapegoat, fault can more easily be technology's, no one's and, in the worst case scenario, common, but it cannot more easily be ours.

Yet, several current technologies, specifically self-tracking technologies in general and the quantified self in particular, seem to focus on single individuals, from their best practices to their worst practices. In what follows, I shall argue that self-tracking technologies and the quantified self cannot be read as a continuation of individual autonomy's rise, but as a symptom of its crisis, in that they sabotage what underpins individual autonomy itself as it has been especially thought of from Kant to Berlin—I shall argue that self-tracking technologies and the quantified self can be read, again, as an overturning of human autonomy.

According to the Oxford English Dictionary, the word "self-tracking" means "The action or practice of monitoring one's own activity, diet, health, etc., now esp. for the purpose of improving one's own well-being and typically by means of recording data in a smartphone".[1] In the last years, the words "quantified self" have been used by the co-editors of "Wired", Wolf and Kelly, to define a phenomenon resulting not only in a website[2] directed by Wolf and collecting self-trackers' practices but also in an academic research institute, i.e. the Quantified Self Institute in Groningen, founded by the Hanze University of Applied Science with the support of the Quantified Self

[1] Oxford English Dictionary, see https://www.oed.com/view/Entry/64799248?rskey=yUDlN3&result=1#eid (accessed in October 2021).

[2] See https://quantifiedself.com/ (accessed in October 2021).

S. Chiodo, *Technology and the Overturning of Human Autonomy*,
Studies in Applied Philosophy, Epistemology and Rational Ethics 66,
https://doi.org/10.1007/978-3-031-26159-6_3

Labs in San Francisco.[3] Speaking of the quantified self means speaking of "self-knowledge through numbers",[4] which is also the subtitle of the website, according to which "The Quantified Self is an international community of users and makers of self-tracking tools who share an interest in 'self-knowledge through numbers'".[5] More precisely, according to Wolf, numbers as self-tracking technologies' outputs are the most reliable knowledge we can obtain. First, we should stop relying on what follows: we "steer by guesswork. We go with our gut" (Wolf, 2010, no page number also in the following quotes) and question our "inner worlds through talking and writing" (Wolf, 2010) to try to find "a truth buried at a deeper level" (Wolf, 2010). But we end up making "errors of fact and errors of judgment" (Wolf, 2010). Second, if we "want to replace the vagaries of intuition with something more reliable" (Wolf, 2010), then we should start relying on what follows: we "first need to gather data" (Wolf, 2010) as "the facts" (Wolf, 2010) we "can live by" (Wolf, 2010). And speaking of data means speaking of numbers: self-trackers "are using numbers" (Wolf, 2010), which "are making their way into the smallest crevices of our lives" (Wolf, 2009, no page number) and "infiltrating the last redoubts of the personal. Sleep, exercise, sex, food, mood, location, alertness, productivity, even spiritual well-being are being tracked and measured" (Wolf, 2010). Finally, "numbers have won fair and square" (Wolf, 2010), which especially means that self-trackers "are constructing a quantified self" (Wolf, 2010).

I shall question the epistemic value of the quantified self in the following sections, which should be introduced by a question on the destiny of the self in technologies of the self, specifically self-tracking technologies. It should be noted that the figure of Prometheus may be embodied by self-trackers themselves, who want not only to improve their health, and even to be possibly resistant to death as a modern Prometheus, but also to create, as a contemporary Prometheus, a kind of further self as a technological exoself that somehow replaces something infinite, i.e. the infinite complexity of the self itself, from its "guesswork" to its "gut" to its "inner worlds" to its "vagaries of intuition". More precisely, the word "exoself" was used by both the founders of the quantified self (see Kelly, 2012) and scholars (see at least Swan, 2013 and Sharon, 2017), who write on Kelly's use of the word "exoself": "the current moment of self-quantification is merely an intermediary step toward something else—the future self. This future self is one that is spatially expanded, with a broad suite of exosenses—the exoself. In fact, self as a concept is a trope that has only arisen recently in the scope of human history, perhaps evolving in lockstep with the sizeable and complex cultures of modernity. The concept of the self continues to shift as individuals react to changes in culture and technology. QS [quantified self] activities are a new means of enabling the constant creation of the self" (Swan, 2013: 95). The notion of self, together with the notion of autonomy strictly correlating with it, has surely experienced a rise in the last three centuries. According to both founders and scholars of the quantified self, the phenomenon

[3] See https://qsinstitute.com/ (accessed in October 2021).

[4] See https://quantifiedself.com/ (accessed in October 2021).

[5] See https://quantifiedself.com/about/what-is-quantified-self/ (accessed October 2021).

of the quantified self itself, specifically the exoself as its consequences, "may have unprecedented regulation and control capabilities. Once equipped with QS devices, an individual body becomes a knowable, calculable, and administrable object. [...] There is a paradox that on the one hand humans are becoming increasingly dependent upon technology for everything including interacting with the outside world, while on the other hand technology is providing a richer, more detailed, controllable, and personal relationship with the world. Ultimately, QS exoself technology is helping to catalyze progress toward a more advanced future society by increasing the most vaunted of human commodities: choice, understanding, consciousness, and freedom" (Swan, 2013: 96). At least two issues are worth noting. First, an idea of knowledge somehow expressed by the words "knowable, calculable, and administrable", which, as it is the case of the epistemic value of the quantified self, I shall question in the following sections. Second, an idea of autonomy somehow expressed by the words "unprecedented regulation and control capabilities" and increasing "choice, understanding, consciousness, and freedom". Thus, self-tracking technologies in general and the quantified self in particular challenge autonomy as follows: on the one hand, as we have seen, current technologies seem to correlate with autonomy's crisis and, on the other hand, self-tracking technologies and the quantified self, as pervasive current technologies, seem to correlate with autonomy's further rise. What is the meaning of the paradox that seems to emerge? From my perspective as a philosopher, the paradox arises precisely from autonomy's crisis itself, in that, as I shall develop in what follows, the idea of autonomy somehow expressed by the words "unprecedented regulation and control capabilities" and increasing "choice, understanding, consciousness, and freedom", i.e. self-tracking technologies' and the quantified self's idea of autonomy, lacks something most essential: an actual νόμος as "law" joining αὐτός as "self"—as I shall develop in what follows, self-tracking technologies and the quantified self are even further symptoms of autonomy's crisis as the overturning of humans' autonomy and technology's automation.

Before proceeding with what follows, I should add at least one last remark. The word "exoself" itself seems to correlate with the word "externalisation" I used to mean the shift from ourselves to technology of nothing less than our most successful definition as humans, i.e. our own core identity. Speaking of exoself seems to lead to a kind of strengthened externalisation, in which several key issues defining humans' core identity are actually externalised: what is more physical ("spatially expanded, with a broad suite of exosenses"), what is more mental ("unprecedented regulation and control capabilities" and increasing "choice, understanding, consciousness, and freedom") and what is physical and mental at the same time ("interacting with the outside world" and "personal relationship with the world"). Thus, if it is true that speaking of exoself leads to a kind of strengthened externalisation, is it also true that speaking of exoself leads to a form of anarchism as our shifting away (ἄν) from what, ruling (ἄρχω) within us without a switch, cannot be switched off by us as we can switch off self-tracking technologies and the quantified self?

3.2 Quantified *je ne sais quoi* from Health to Happiness

From a philosophical perspective, the words "guesswork", "gut", "inner worlds" and "vagaries of intuition" correlate with the notion of *je ne sais quoi* (see at least Kohler, 1953–54 and D'Angelo & Velotti, 1997), which was developed between the seventeenth century and the eighteenth century especially starting with Descartes' (who acknowledges a sphere far from clarity and distinction, see Descartes, 1644) and Leibniz's (who continues to acknowledge a sphere of obscurity and confusion, see Leibniz, 1684 and 1704) philosophical works, according to which the domain of *je ne sais quoi* is somehow complementary to the domain of λόγος (transliterated as *logos*), whose first meaning is, most interestingly, "computation, reckoning"[6] (I shall go back to the notion of *logos* in what follows). Bouhours, who first introduced the notion of *je ne sais quoi* (see Bouhours, 1671), defines it as something definitely complementary to the notion of *logos*, i.e. as "l'asyle de l'ignorance", "car il me semble qu'on se trouve toujours là, quand on ne sçait plus que dire" (Bouhours, 1671: 344). Thus, there is even a kind of opposition between what falls under the domain of *je ne sais quoi* and what falls under the domain of *logos*: the former is not only acknowledged but also thought of as something that cannot be "comput[ed], reckon[ed]", in that it exceeds the latter—yet, the domain of *je ne sais quoi* is thought of as one of the domains of the human mind (for instance, we may happen to say that the best decision we have ever made was made by going "with our gut", which is precisely what the quantified self wants to replace with "comput[ing], reckoning" self-tracking technologies' "numbers").

Interestingly enough, between the seventeenth century and the eighteenth century, European culture gives rise not only to *je ne sais quoi* but also, and especially, to the first scientific revolution (as what is standardly placed between the sixteenth century, with Copernicus' work, and the eighteenth century, with Newton's work) as a further, and strongest, rise of *logos*: the former domain arises not as an outcome of an era characterised by irrationality, but as a most rational era's outcome, and Descartes and Leibniz themselves make greatest contributions to both the first scientific revolution and *je ne sais quoi*, which somehow means that at least two domains have to be acknowledged, i.e. the domain of *logos* where particulars can be explained through universals and the domain of *je ne sais quoi* where particulars characterised by a kind of "computation[al], reckon[able]" irreducibility cannot be explained through universals.

The domain of the latter kind of particulars, which are somehow residual, but pervasive in human life, is precisely what the quantified self wants to challenge. Self-trackers' testimonies published in the website of the quantified self are most interesting to understand the quantified self's attitude. After having read several

[6] Liddell-Scott-Jones Greek-English Lexicon, see https://artflsrv03.uchicago.edu/philologic4/LSJ/query?report=bibliography&method=proxy&head=logos&start=0&end=0 (accessed in October 2021).

tens of testimonies collected in the section "Show & Tell",[7] where both videos and transcripts are published, I selected two testimonies as most representative cases in point to analyse: the first one is on health, entitled *Where There's Data There's Hope*,[8] and the second one is on happiness, entitled *Happsee*.[9] In what follows, first, I shall review them and, second, I shall argue that they, as most representative cases in point of the current relationship between humans and technology, can illuminate at least two crucial phenomena: a kind of radical particularisation that may be read, again, as anarchism and a kind of reshaping of humans' core identity that may be read, again, as a trade of autonomy for automation.

The first testimony of self-tracking, *Where There's Data There's Hope*, is offered by a professor of computer science and engineering who self-diagnosed Crohn's disease before his doctor. He used self-tracking technologies to be "healthier". When he asked his doctor to take blood tests to have further details on his health, his doctor answered that there was "no insurance code for that because it's preventative". Thus, he used the company Your Future Health[10]: "we now know it's something like 25% of all blood tests in the US are not associated with hospitals or doctors. So I said, wow this is cool. What else can I learn?". And he learned that companies can do almost whatever can be paid. After having taken blood tests, an inflammation emerged as his health problem to solve. He asked: "what in the world could be causing this inflammation inside of me"? And he answered: "so because I had got out of the doctor thinking and hospital thinking and into the web thinking, while I was at yourfuturehealth.com, [I asked] what other tests could I do". Thus, he continued to take tests, and after several tests an inflammatory bowel disease (IBD) emerged as his health problem to solve, which his doctor did not considered. The opposition between what emerged through both self-tracking and self-managing and what did not emerge through going to his doctor led him to the following words: "my doctor who had done the colonoscopy had said you don't have IBD [...]. You know I've done your colonoscopy, I've been inside and you don't have it. And I said [...] you must be doing these all day long. He said, yeah I do, I do dozens of these a day. I said, so that's why you don't have time to read the scientific literature". He also used the company 23andMe[11] as follows: "you can just go to 23andMe and you type in Crohn's and up comes these [...] points along your DNA out of the 1 million that they have on 23andMe, which have been shown to be associated with people who have IBD or Crohn's. And on some of them I'm green and I'm less likely to have

[7] See, also for the following quote, https://quantifiedself.com/show-and-tell/ (accessed in October 2021). Self-trackers' testimonies are collected as follows: "chronic condition, cognition, diet and weight loss, environment, food tracking, genome and microbiome, heart rate and cardiovascular, location, media, metabolism, money, mood and emotion, other, ovulatory cycle and pregnancy, productivity, sleep, social life and social media, sports and fitness, stress".

[8] See, also for the following quotes, https://quantifiedself.com/show-and-tell/?project=390 (accessed in October 2021).

[9] See, also for the following quotes, https://quantifiedself.com/show-and-tell/?project=723 (accessed in October 2021).

[10] See https://www.yourfuturehealth.com/ (accessed in October 2021).

[11] See https://www.23andme.com/ (accessed in October 2021).

it, but this one is Interluken-23 Receptor [...]. It's a pro-inflammatory chemical that basically I make more than the average person. So does that have anything to do with Crohn's? So I look up the Intreluken-23 Receptor, and sure enough here it is, Interluken-23 is the master regulator in Crohn's disease. So I'm self-teaching myself this as I'm taking the data". Finally, he self-diagnosed Crohn's disease.

The self-tracker's further words disclose his epistemological attitude, which is analogous to Wolf's. On the one hand, he attacks the domain of *je ne sais quoi*: "this idea that you can feel what's going on inside of you [...] is just so epistemologically false; you just can't do it. And so as you'll see, there are things that are completely out of whack inside me and I would never know that by 'feeling', which is the first thing your doctor asks you when you come in the office. [...] So what I've learned by this experience over 10 to 12 years is [...] [that] when you go into the doctor's office and he says how do you feel, this is next to useless". On the other hand, he defends the domain of *logos*: "don't trust the diet book, actually measure", in that "What you want to say is what are your numbers, what do they look like over time, how do they compare with the population. And if that's what the doctor looks at [...] then you can have a useful conversation. So my endpoint where I've come to is where there's data, where you can actually quantify your body and get a sense of knowledge and a sense of if not control and an ability to at least understand what's happening to you there's hope". It is worth noting that, in addition to "the web", "scientific literature" and "Wikipedia" are the ways in which he "tak[es] the data": "I had no idea what lactoferrin was. I had to self-teach myself by going to scientific literature and using you know what Wikipedia say lactoferrin is".

The first testimony's key points are at least three, which I shall develop in the following sections. First, the epistemological move from the expert, as the mediator between knowledge and the nonexpert, to a kind of do-it-yourself expertise (which is especially expressed in terms of getting "out of the doctor thinking and hospital thinking and into the web thinking", in addition to "scientific literature" and "Wikipedia", which are self-reviewed: "I'm self-teaching myself this as I'm taking the data"). Second, the epistemological move from the domain of *je ne sais quoi* to the domain of *logos* (which is especially expressed in terms of replacing "feeling" "what's going on inside of you" with "measur[ing]" "what are your numbers"). Third, the socio-cultural and, finally, political move from public healthcare to private healthcare (which is especially expressed in terms of turning to companies: if there is "no insurance code for that because it's preventative", then "it's something like 25% of all blood tests in the US are not associated with hospitals or doctors". I shall not develop the issue of the move from public healthcare to private healthcare in what follows, which also correlates with reshaping the difference between the meaning of being healthy and the meaning of not being healthy, but see at least Dickenson, 2013).

The second testimony of self-tracking, *Happsee*, is offered by an individual who is both the creator and the user of an app based on smartphone sensors that, through self-tracking, collect his data to quantify his happiness. He created his app after having decided, first, "to measure his happiness and mood to improve it" and, second, that other self-tracking technologies "did not meet his needs". He started with something interesting from an epistemological perspective: "I was in a situation where I thought I was unhappy. So I thought I was in this situation where I thought I was unhappy [...]. [But] I didn't know if I was unhappy". What is somehow internal is considered as uncertain, but what is somehow external, specifically technology, is considered as certain and, consequently, the former, specifically happiness, is somehow externalised by the latter, specifically "this android app called Happsee": "So I wanted to explore this whole space of unhappiness and how to improve it. So how do we do that? I wanted to measure my happiness and mood", "So I looked at a lot of measurements tools. I looked at a lot of android apps. I looked at a lot of web apps. Most of them did not suit my needs. They were either too complex. They asked me to fill out a bunch of surveys and I think a few of them actually made me angry, because I was like [asking myself] why am I filling out all these questions and that completely defeats the purpose of understanding whether I'm happy. [...] So in the spirit of software I made my own tool". Happsee worked as follows: "I entered in some moods. I entered in how happy I was on a zero to ten scale. I could type some stuff in if I wanted and I could attach a photo. And using that it could generate a bunch of graphs". Even though, paradoxically enough, he "entered in" precisely what was somehow internal, i.e. "some moods" and "how happy I was" by translating them into "a zero to ten scale", he was quite enthusiastic: "So uncertain is dropping. Stressed was in don't know, I was never really that stressed, and bored is dropping, so I'm not bored at all now. So negative moods have been changing and my positive moods have also been changing and they've also been going up. So I'm more energetic which is this line down here, I'm more relaxed, more engaged and I am also more confident I guess maybe".

The self-tracker's further words disclose his epistemological attitude, which adds something interesting to the first self-tracker's. First, knowledge and obviousness seem to converge: "being tired had a huge effect on my happiness, and when I started to get less tired, I started to get a lot happier". Yet, enthusiasm led him to say: "So what am I thinking for the future? So I want to use machine learning to predict happiness". Second, knowledge, which seems to result from a kind of particularity that seems to converge with his own idiosyncrasies, starting with his own laziness, seems to result in a kind of tailored-made knowledge: "I'm lazy, and the one thing that gets me annoyed a little bit is [that] sometimes I have to make a lot of entries to tell the system how happy I am. What if it could just pick that up from passive data and I think that's one thing I want to explore more" "to maximize information gathered while minimizing the information I actually had to type in because I'm lazy".

The second testimony's key points are at least six, which I shall develop in the following sections. First, again, the epistemological move from the expert, as the mediator between knowledge and the nonexpert, to a kind of do-it-yourself expertise (which is especially expressed in terms of replacing experts in happiness as, for instance, psychiatrists and psychologists with the "android app called Happsee"). Second, again, the epistemological move from the domain of *je ne sais quoi* to the domain of *logos* (which is especially expressed in terms of replacing thought, specifically uncertainty resulting from "I thought I was in this situation where I thought I was unhappy", with "measurements tools"). Third, again, the socio-cultural and, finally, political move from public healthcare to private healthcare (which is especially expressed in terms of turning to no company, but even to himself: "I looked at a lot of measurements tools", but "Most of them did not suit my needs", "So in the spirit of software I made my own tool"). In addition to the key points shared by the two testimonies, at least further three issues are worth noting, which I shall develop in the following sections. First, a kind of externalisation emerges. There is an opposition between what he thinks (which is uncertain: "I was in a situation where I thought I was unhappy. So I thought I was in this situation where I thought I was unhappy") and what he knows (which is certain: "[But] I didn't know if I was unhappy"). And moving from thought's uncertainty to knowledge's certainty means moving from what is internal, i.e. "some moods" and "how happy I was", to what is external, i.e. "this android app called Happsee". Second, a convergence between knowledge and obviousness emerges: even though it is quite obvious that the less tired one is, the happier one is, the output according to which "when I started to get less tired, I started to get a lot happier" causes enthusiasm. Third, a convergence between knowledge and a kind of idiosyncratic particularity emerges: the former seems to move from being underpinned by universality to being underpinned by the latter (which is especially his laziness). Thus, knowledge seems to move from being something that applies in a universal way to being something that applies in a particular way, and even in a tailored-made way.

Finally, the self-tracker's words "So in the spirit of software I made my own tool" should be stressed, in that they disclose an attitude that is both positive from a practical perspective (if I have a problem, then I can fix it without turning to others) and challenging from an epistemological perspective (if I have a problem, then I can fix it without turning to experts, from psychiatrists and psychologists to the best computer scientists, in that tailored-made knowledge is the best knowledge, even better than universal knowledge as, at least, shared knowledge). In addition, "the spirit of software" seems to imply that almost anything is a problem to fix—conversely, "the spirit of" philosophy implies that sometimes fixed problems are worse than unfixed problems and sometimes problems to fix are not to fix at all, in that they are not problems at all, but opportunities to seize.

Before proceeding with the following sections, I should add that the two self-trackers share a positive attitude from a practical perspective: both of them want to proactively improve, from the first self-tracker who wants to be "healthier" to the second self-tracker who wants "to improve" his happiness. In what follows, I shall exclusively question the epistemological meaning of their attitude for at least two key reasons. First, humans' actions, which can impact several others' lives, arise from humans' epistemological attitudes—what if I take nothing but a correlation underpinned by nothing but tailored-made knowledge instead of knowledge that, going back to Reichenbach's Hamlet, "giv[es] me the courage I need for my action"? Second, the two self-trackers' attitude is not only shared by both other self-trackers and several others who are not regular self-trackers but also, and especially, symptomatic of the philosophical meanings technology is increasingly taking on, from technology's key role played in addressing the divine's destiny to technology's key role played in addressing human autonomy's destiny and, consequently, human identity and dignity.

3.3 Radical Particularisation and Anarchism: Crisis of Ideal Models?

Two key points are explicitly shared by the two testimonies: first, the epistemological move from the expert, as the mediator between knowledge and the nonexpert, to a kind of do-it-yourself expertise and, second, the epistemological move from the domain of *je ne sais quoi* to the domain of *logos* (in addition to the issue of the move from public healthcare to private healthcare I shall not develop in what follows). In addition, at least two key points are implicitly shared by the two testimonies: first, a kind of externalisation from (internal) thought's uncertainty to (external) knowledge's certainty (which, in the case of the first testimony, can be found in the move from the question on what your feelings are to the question on what your numbers are) and, second, the convergence between knowledge and a kind of idiosyncratic particularity, according to which knowledge seems to move from being something that applies in a universal way to being something that applies in a particular way, and even in a tailored-made way (which, in the case of the first testimony, can be found in the move from public healthcare to 23andMe).

As far as the convergence between knowledge and obviousness is concerned, even though it does not emerge in the case of the first testimony, it is explicit in several self-trackers' testimonies. For instance, according to the testimony entitled *My Life in 40 Variables*,[12] happiness is found to be correlated with a series of obvious things: "this is a chart of my happiness over a year, and the first thing I find out [...] [is that] it's a little bit higher [...] [on] the days that I've been on holiday. So it's pretty good to know that [when] I'm on holiday I'm happy. But I don't really want that in my database because it was different from my normal life. So I extracted them and asked my computer well, what makes me happy and he says well, you have to be healthy, just be active, have a good night sleep and don't be stressed. And that's pretty nice to know that actually what you suppose what you think will make you happy does make you happy". Interestingly enough, obviousness (for instance, the correlation between happiness, on the one hand, and the first testimony's being "less tired" and the second testimony's being on "holiday", "healthy", "active", having "a good night sleep" and not being "stressed", on the other hand) makes self-trackers quite enthusiastic, which may be a meaningful symptom of an epistemological attitude: what is self-acquired is satisfactory to the point that even its obviousness replaces a more meaningful knowledge. But what is the reason why what is self-acquired is satisfactory even though it is nothing but obviousness? From an epistemological perspective, a hypothesis that several testimonies seem to imply is the following: what is self-acquired is easier,

[12] See, also for the following quote, https://quantifiedself.com/show-and-tell/?project=847 (accessed in October 2021).

in that actual contradicting comparisons do not emerge, and what is easier is self-reassuring—and, if it is true that our technological era may be read as our attempt to unburden ourselves of our unbearable burdens, it is also true that, from an epistemological perspective, thinking of knowledge as something easy, specifically self-acquirable and self-reassuring, can surely help us face Hamlet's unbearable questions "But shall I be right in this case?", "How then can I act in this case?". And the kind of easy knowledge that emerges as something self-acquirable and self-reassuring correlates, again, with epistemological anarchism: speaking of self-acquirable and self-reassuring knowledge as something easy, in that it is not challenged by actual contradicting comparisons, means speaking of shifting away (ἀν) from what rules (ἄρχω) both as the expert (from doctors to psychiatrists and psychologists to the best computer scientists) and, especially, as a shared method of knowledge—speaking of easy knowledge, specifically self-acquirable and self-reassuring, correlates with epistemological anarchism, in that what rules as a universal method is shifted away by a self-referential, and even idiosyncratic, particularity.

The last outcome goes back to two of the key points described above, specifically the epistemological move from the expert to a kind of do-it-yourself expertise and the epistemological move from knowledge as something that applies in a universal way to knowledge as something that applies in a particular way, and even in a tailored-made way (as far as the other two key issues are concerned, I shall develop them in the following section and chapters). Other testimonies are worth quoting. A first self-tracker attacked his doctors in his testimony entitled *Debugging My Allergy*: "They [my doctors] can't help me, so I'll try and find these methods of my own health condition",[13] which means starting self-tracking. And a second self-tracker, celebrated by Wolf "for being disloyal to the professional, institutional version of science, for not conforming to scientific rituals" (Wolf, 2010, also reported in Sharon, 2017: 110), explained, in his testimony entitled *QS + Paleo = ?*,[14] that he started eating "half a stick of butter a day", in that, according to his self-tracking technologies, "butter makes my brain work better. [...] I really benefit from this and perhaps other people would too". After his death at the age of sixty because of coronary occlusion, his mother wrote in his blog: "Most of you won't be surprised to learn that Seth had not visited his doctor in Berkeley in many years".[15] From a sociological perspective, the phenomenon stressed by the last two testimonies is readable both as a rejection of experts in particular (for instance, according to Lupton, "patients can and should become experts on themselves with the help of these technologies. [...] this will radically alter the doctor-patient relationship, shifting the balance of power from doctors as medical experts to patients as more knowledgeable than ever about their own bodies thanks to the affordances of digital technologies that allow them to generate personal health information", Lupton, 2016: 78) and as a rejection of rules in general (for instance, according to Heyen, the quantified self is "a means of

[13] See https://quantifiedself.com/show-and-tell/?project=563 (accessed in October 2021).

[14] See, also for the following quote, https://quantifiedself.com/show-and-tell/?project=206 (accessed in October 2021).

[15] See https://sethroberts.net/category/general/ (accessed in October 2021).

resistance against and a remaking of dominant social norms and conventions", Heyen, 2020: 1699). It should be added that, according to the founders of the quantified self, the phenomenon described above is readable as "personal science"[16] (Wolf & De Groot, 2020: 1), which "involves the deliberate choice of the individual about what questions to ask, what methods to use and what observations to make" (Wolf & De Groot, 2020: 2). The website of the quantified self also introduces the personal dimension: "The purpose of this website is to support anybody who wants to use empirical methods to explore personal questions. We call this practice 'everyday science'",[17] which has to do with "using their [self-trackers'] own data", and the website aims at "making it easy to get started". More precisely, the founders of the quantified self "envision a world of personal scientists" (Wolf & De Groot, 2020: 4), in which, differently from what happens in the case of citizen science and in the case of N-of-1 trials, single individuals play at the same time whatever role. First, they are the investigators: "non-professionals occupy most, if not all, of the significant roles in research [...] [that] is self-directed: the subject of the research is also the primary investigator" (Wolf and De Groot 2020: 4). Second, they are the investigated: "The selection of topics and questions [...] are determined by the researcher's personal motive alone" (Wolf and De Groot 2020: 4), who chooses not only "highly individual [...] personal challenges" (Wolf and De Groot 2020: 4 and 3) but also "measurements that are personally relevant rather than clinically defined" (Wolf and De Groot 2020: 3–4). Third, they are the users of the investigation: "the discoveries are applicable directly by the person doing the research" (Wolf and De Groot 2020: 4). Analogously, according to the website of the Quantified Self Institute, the quantified self "is the gathering of personal data for and by you. [...] It is a functionally 'selfish' activity, which is the result of a personal motivation. 'Me and my data', that is the point with Quantified Self".[18] And, if the objection is that the collection of self-trackers' testimonies is a way to move from the personal dimension to the common dimension, then sociologists frequently question its success: on the one hand, "public presentations of self-tracking research at meetups or conferences make hardly any references to other self-trackers and their activities" (Heyen 2020: 133) and, on the other hand, if it is true that "the self-tracker can only reproduce a research study on himself or herself, because here the researcher and the test subject are one and the same person" (Heyen 2020: 131), it is also true that "objectivity in the sense of intersubjective verifiability or reproducibility of the measurement plays almost no role in the meetup discussions" (Heyen 2020: 131).

---

[16] The notion of personal science arises from Polanyi's stress on the role played by subjectivity in scientific practice, from discovery to validation (see Polanyi, 1958). See also other scholars' reflections upon the relationship between knower and knowledge, with a recent stress on the quantified self, especially Heyen (2020), in addition to Martin and Brouwer (1993) and De Groot et al. (2017).

[17] See, also for the following quotes, https://quantifiedself.com/get-started/ (accessed in October 2021).

[18] See http://qsinstitute.com/quantified-self-quantified-us-quantified-other/ (accessed in October 2021).

From a philosophical perspective, the issue at stake is the meaning of the kind of radical particularisation that arises from the solipsistic coincidence between investigator, investigated and user of the investigation. The coincidence described above makes nonexperts, i.e. a great number of single individuals, capable of "practic[ing] 'everyday science'", in that "get[ting] started" is "easy" if what single individuals need is nothing but self-tracking technologies, and even nothing but smartphones, and no expertise at all. But the easiness of "practic[ing] 'everyday science'" should be questioned on science's easiness itself: what is the epistemic value of a kind of science that is easy precisely because of the removal of the most difficult thing, i.e. expertise? If there is no expertise, then what is the epistemic value of data as inputs? And, if there is no expertise, then what is the epistemic value of data as outputs? The questions are more urgent if we consider that, from an epistemological perspective, it has been acknowledged for decades that speaking of data does not mean speaking of something neutral (see at least Ihde 1979, 1993, 2001 and 2009 and Latour 1987, 1993 and 1999)—conversely, speaking of data means speaking of something requiring both expertise and contexts, specifically the expertise in contextualising data. The meaning of eating "half a stick of butter a day" cannot be absolutised and reduced to one's "brain": it needs both contexts, specifically one's general health conditions, and, especially, expertise in reading the relationship between contexts and eating "half a stick of butter a day", specifically doctors' expertise. And the meaning of "some moods" cannot be absolutised and reduced to one's "happiness": it needs both contexts, specifically one's general psychological condition, and, especially, expertise in reading the relationship between contexts and "some moods", specifically psychiatrists' and psychologists' expertise. Thus, what is the meaning of the exponential frequency of use of self-tracking technologies, if it has been acknowledged for decades that speaking of data means speaking of something that requires both expertise and contexts?

I believe that a promising way to read the phenomenon described above is epistemological anarchism. Whenever the solipsistic coincidence between investigator, investigated and user of the investigation is added to the removal of expertise, we face the questionable do-it-yourself expertise leading to the equally questionable tailored-made knowledge we have seen, which is caused by the following epistemological phenomenon. First, we shift away (ἄν) from what rules (ἄρχω) by skipping our expertise, in that we rely on our data reading even though we are nonexpert. Second, we shift away (ἄν) from what rules (ἄρχω) by skipping experts' expertise, in that we do not rely, for instance, on doctors', psychiatrists', psychologists' and the best computer scientists' data reading even though they are experts.

And, if the objection is that there is still something that rules, i.e. data, then the answer is that, again, data cannot rule without sense-making—technology cannot rule without sense-making, and what can make sense is human, at least in the sense that data as technology's output cannot do without ending up meaning something, whatever it is, to humans, from their past to their present to their future, both when data can mean something to a single individual whose existence has something to do with it, directly or indirectly, and when data can mean something to a society whose existence has something to do with it, directly or indirectly (as most philosophies, also limiting ourselves to the last century, argue, from hermeneutics to constructivism). If it is true that what we actually want to know is what "half a stick of butter a day" means to a human being's life and what "some moods" mean to a human being's life, it is also true that what we actually want to know cannot do without humans' sense-making (which requires both expertise and contexts)—thus, technology alone cannot make sense, which means that technology alone cannot (ἀν) rule (ἄρχω): technology without humans' sense-making, specifically experts' sense-making, means epistemological anarchism.

The words of the founders of the quantified self seem to show even a form of political anarchism added to a form of epistemological anarchism. As we have seen, a self-tracker is celebrated by Wolf "for being disloyal to the professional, institutional version of science, for not conforming to scientific rituals". In addition, he writes that the "idea that we can – and should – defend ourselves against the imposed generalities of official knowledge is typical of pioneering self-trackers" (Wolf 2010) he celebrates. Thus, it is not only a matter of removing the ruler as the scientific method (as it is thought of especially starting with Galilei's massive use of idealisation together with experimentation, even though differences and counterarguments emerge), together with expertise, experts and the public comparisons between their outcomes, but also a matter of removing the ruler as what is "institutional", "ritual" and "official". It is no coincidence that scholars agree on the correlation between self-tracking technologies, especially the quantified self, and a kind of radical freedom, when they report a self-tracker's words: "'tracking your weight yourself and having a doctor put you on a scale are not the same'. For her, the choice to actively track herself is, as she put it, 'liberating'" (Sharon and Zandbergen 2017: 1702). A kind of escape is also reported (see Nafus and Sherman 2014).

Before questioning, in the following section, the paradoxical relationship between anarchism and freedom (if we consider Kant's idea according to which autonomy, and not anarchism, underpins freedom), we should reflect upon another issue: the solipsistic coincidence between investigator, investigated and user of the investigation, added to the removal of comparisons both in terms of experts' outcomes and in terms of what is "institutional", "ritual" and "official", leads to a kind of individualism (Morozov even talks about exceptionalism, see Morozov 2013) that is not only epistemological but also social, and even existential. As far as the epistemological dimension is concerned, what is at stake is, again, the expertise's destiny also in terms of its more general transformation (see at least Collins 2014 and Collins and Evans 2002 and 2007). As far as the social, and even existential, dimension is concerned, what is at stake is the human sociality's destiny: if it is true that "the idea that what is 'good', 'right', or 'healthy' for one person differs for every individual is a fundamental axiom" (Sharon 2017: 109), it is also true that, as a self-tracker says, "you know yourself what is good for you or what is not good for you" (reported by Heyen 2020: 135), which means not only epistemological solipsism but also socio-existential solipsism, in that skipping comparisons implies skipping both relationships between our data and others' data and relationships between us and others. But individualism may be reassuring precisely by implying skipping comparisons, in that comparisons may be contradicting and, consequently, challenging—contradicting comparisons may be another kind of burden of which we try to unburden ourselves through technology.

The opposition between the solipsistic dimension meant by self-tracking technologies and the quantified self and the public dimension meant by the scientific method leads to a further opposition. On the one hand, we have what, being solipsistic, can be nothing but particular and, on the other hand, we have what, being public, can be universal. If "the idea that what is 'good', 'right', or 'healthy' for one person differs for every individual is a fundamental axiom", then what is the destiny of the scientific method? More precisely, what is the destiny of modern and contemporary science, together with its epistemological roots in ancient culture, underpinned by a precise move, which is the move from particularity to universality through abstraction and idealisation[19]?

Self-tracking technologies and the quantified self are not sufficient to argue that the cornerstones of both science in particular and knowledge in general, i.e. ideal models resulting from the move from particularity to universality through abstraction and idealisation, experience a crisis. Yet, self-tracking technologies and the quantified self as clear symptoms of a more general use of technology and, consequently, of a more general epistemological attitude, can illuminate ideal models' destiny. Not only whenever we use our smartphones for self-tracking but also whenever we use Google for self-diagnosis, and even whenever we use social networks for posting whatever on whatever, we exercise a form of epistemological anarchism that increasingly undermines ideal models resulting from the move from particularity to universality through abstraction and idealisation. For instance, if I, as a nonexpert, post in a matter of minutes that a first given individual is wrong about a first given issue, a second given individual is wrong about a second given issue and a third given individual is right about a third given issue, then I exercise a form of epistemological anarchism, in that my posts are not (ἄν) underpinned by expertise as what can actually rule (ἄρχω). Yet, I do not consider them as (legitimate) private opinions, i.e. δόξα (transliterated as *doxa*) meaningfully opposed to ἐπιστήμη (transliterated as *episteme*) as knowledge, specifically scientific knowledge. Conversely, whenever I post them on social networks, I make them move from a (legitimate) private dimension to a (questionable) public dimension, which means that I (questionably) confuse the two dimensions: whenever I make them move from what is private to what is public, I actually make δόξα occupy the dimension of ἐπιστήμη—whenever I make them move from what is private to what is public, I actually make δόξα replace ἐπιστήμη, which means that I actually exercise a form of epistemological anarchism, in that I actually shift away (ἄν) what can rule (ἄρχω), i.e. ἐπιστήμη, to replace it with what cannot rule, i.e. δόξα.

---

19 See especially Brzeziński et al. (1990a, 1990b), Brzeziński and Nowak (1992), Kuokkanen (1994), Nowakowa and Nowak (2000), Coniglione et al. (2004), Cartwright and Jones (2005) and Brzechczyn (2009). See also Huxely (1938), Rescher (1987), Berlin (1988), Emmet (1994), Jullien (2009) and Ilyenkov (2012).

And the form of epistemological anarchism I exercise increasingly undermines ideal models. When I keep my opinions private, i.e. when I do not use social networks for posting whatever on whatever, I implicitly continue to think of knowledge as something requiring the following process: I have a given particular and, if I want to judge it correctly, then I should refer it to a more general context, which should be even universal from an ideal perspective. For instance, if I want to judge a particular work of art correctly, then I should refer it to the ideal work of art, which means that I should answer the following questions: what should a work of art be? More precisely, what should any work of art be? What qualities should a work of art possess to be actually defined as a work of art? More precisely, what qualities should any work of art possess to be actually defined as a work of art? Finally, I should answer the following question: is the particular work of art I want to judge eligible to be actually defined as a work of art, and why? And, if I want to judge a particular vaccine correctly, then I should refer it to the ideal vaccine, which means that I should answer the following questions: what should a vaccine be, specifically to be safe and effective? More precisely, what should any vaccine be, specifically to be safe and effective? What qualities should a vaccine possess, specifically to be safe and effective? More precisely, what qualities should any vaccine possess, specifically to be safe and effective? Finally, I should answer the following question: is the particular vaccine I want to judge eligible to be actually defined as a vaccine, specifically safe and effective, and why? If I do not have any artistic expertise, then I cannot answer the first series of questions. And, if I do not have any medical expertise, then I cannot answer the second series of questions. I can express δόξα, i.e. opinions, but not knowledge, i.e. ἐπιστήμη. Thus, I keep my opinions private not to contribute to confuse knowledge with opinions. Conversely, when I do not keep my opinions private, i.e. when I use social networks for posting whatever on whatever, I implicitly stop thinking of knowledge as something requiring the process described above: I implicitly think that, if I want to judge a given particular correctly, then I should not refer it to a more general context, which should be even universal from an ideal perspective. More precisely, I implicitly think that, if I want to judge a given particular correctly, then I should express something equally particular, i.e. my particular opinion. For instance, I express my particular opinion on a particular work of art without any artistic expertise. And I express my particular opinion on a particular vaccine without any medical expertise. Moreover, I do not keep my opinions private, but I post them on social networks, in that, if it is implicitly true that there are only particulars, and not universals, starting with ideal models, it is also implicitly true that I do not contribute to confuse knowledge with opinions: if the latter occupy the dimension of the former, which is removed, then knowledge is opinions. And I am an epistemological anarchist,[20] in that I remove (ἀν) what

[20] Feyerabend's description of his epistemological anarchist, when he argues against the kind of normativity that the scientific method requires, is fascinating: "the one thing he opposes positively and absolutely are universal standards, universal laws, universal ideas such as 'Truth', 'Reason', 'Justice', 'Love', and the behaviour they bring along" (Feyerabend, 1975: 189), in that "there is only one principle that can be defended under all circumstances and in all stages of human development. It is the principle: anything goes" (Feyerabend, 1975: 28), according to which "'Truth', written 'in capital letters,' is an orphan in this world, without power and influence" (Feyerabend, 1987:

can rule (ἄρχω), as knowledge underpinned by expertise in referring particulars to universals, starting with ideal models, to replace it with what cannot rule.

Thus, the question on the ideal models' possible destiny seems quite urgent in an era characterised by an impressively increasing use of technology as what can make us skip expertise as the difficult capacity to move from particularity to universality through abstraction and idealisation. Becoming epistemological anarchists who "practice 'everyday science'", which "mak[es] it easy to get started", is far easier than becoming experts who exercise the difficult capacity to refer particulars to ideal models.

And whenever we, as epistemological anarchists, use technology to make it easier and easier to shift away (ἄν) from the difficult burdens implied by what can rule (ἄρχω), we equally shift away from the greatest ruler: the law (νόμος) we give ourselves (αὐτός), and which finally makes us free—whenever we use technology to make it easier and easier to shift away from the difficult burdens implied by what can rule, we equally shift away from our autonomy, which finally makes us free. Even though what we may actually want to obtain is precisely to unburden ourselves of the burdens of both rulers in general and our autonomy in particular, we should ask ourselves if humans who trade their autonomy for technology's automation are also what we may actually want to obtain.

3.4 Reshaping Humans' Core Identity: Trading Autonomy for Automation

Yet, surprisingly enough, freedom happens to be thought of as correlated more with anarchism than with autonomy. For instance, as we have seen, "'tracking your weight yourself and having a doctor put you on a scale are not the same'. For her, the choice to actively track herself is, as she put it, 'liberating'". But from what do we want to free ourselves? If my argument on anarchism makes sense, then the possible answer is that we want to free ourselves precisely from what anarchism makes us shift away: both rulers (ἄρχω) in general, starting with others' expertise, and our autonomy (νόμος) in particular, starting with our expertise in moving from particularity to universality through abstraction and idealisation, which finally underpin our decisions and actions. And why do we want to free ourselves from both rulers in general and our autonomy in particular? As we have seen in several cases, the possible answer is that our key reason is a matter of unbearable burdens, both epistemological and ethical, of which we want to unburden ourselves through technology as our perfect scapegoat.

---

102). "[L]ike the Dadaist, [...] he 'not only has no programme, [he is] against all programmes'" (Feyerabend, 1975: 189) and "becomes capable of stepping outside the most fundamental categories and convictions, including those which allegedly make him human" (Feyerabend, 1975: 189. See also Rorty, 1982, 1989).

Self-tracking technologies and the quantified self give us further clues to reflect upon the key reason why we want to free ourselves from both rulers in general and our autonomy in particular. A first clue has to do with something we may correlate with autonomy's irresistible rise, which seems to be followed by autonomy's equally irresistible crisis. As Lupton remarks in the case of a columnist of "The New York Times", "he engaged in self-tracking because he believed that '[y]ou want to be your best self'" (Lupton, 2016: 65). And being "your best self" means "behaving as responsible citizens, engaged in self-care [...], engaging in self-optimisation or enhancement [that is] is even demanded of people" (Lupton, 2016: 48). But, as the quantified self shows, after self-tracking to pursue "ideal selfhood" (Lupton, 2016: 50), which correlates with autonomy through the demand of being "responsible" both as single individuals "engaging in self-optimisation or enhancement" and as "citizens", self-trackers frequently experience a crisis, and even a breakdown. The following case is expressive: "Yes, I did it. / On a crisp Tuesday morning / after 40 measurements a day for 1,5 years / I. Stopped. Tracking. / Why? / [...] I had stopped trusting myself / letting the numbers drown out / my intuition / my instincts[.] / Each day / my self-worth was tied to the data[.] / One pound heavier this morning? / You're fat. / 2 g too much fat ingested? / You're out of control. / Skipped a day of running? / You're lazy" (Carmichael, 2010, no page number). The case quoted above is also interesting for a second reason I shall develop in the following chapters: the destiny of human identity if humans totally replace feeling their selves through "intuition" and "instincts" with computing their selves through "measurements", "numbers" and "data". But the first reason why the case quoted above is interesting is that it clearly shows that there is a fine line between autonomy's rise and autonomy's crisis. The more humans' core identity is defined as autonomous, the more autonomy is demanded of humans even as individual responsibility for "self-optimisation or enhancement" by self-tracking through "40 measurements a day". And the more individual responsibility by self-tracking is demanded of humans, the heavier, and even unbearable, the autonomy's burden humans have to bear is. Finally, humans stop. But stopping does not necessarily mean stopping tracking. Conversely, stopping seems to frequently mean, more artfully, using technologies, from self-tracking technologies to others, to win over the real enemy, which is not technology at all, but autonomy—and what is most artful is that technology, precisely as Prometheus' best "gift to man" told by Plato, is far from being humans' real enemy: technology is humans' best ally, and even humans' best scapegoat.

A second clue has to do with something analogous. A self-tracker writes: "We (the apps and I) had co-constructed a digital model of my self, and here I was, managing myself, it seems, by proxy. The feedback from that digital model often took precedence over how I physically felt. When I didn't eat 'enough' protein I felt weaker, and when I had too much sugar I felt fatter. These were delayed reactions; a re-reading of my body from the model. I've yet to decide: is this model pushing me closer in contact or further away from my self and my world?" (Williams, 2013: 3). Again, we can find the issue I shall develop in the following chapters, i.e. the destiny of human identity if humans totally replace "I" and "how I physically felt" with "the apps" and "The feedback from that digital model". And we can also find the following issue, which gives us insights to reflect upon the key reason why we want to free ourselves from both rulers in general and our autonomy in particular: the self-tracker uses technology as "a digital model of my self", as a "proxy". His question is "is this model pushing me closer in contact or further away from my self and my world?". Our possible answer is that what we may actually want to obtain is precisely to be "push[ed]" "further away from my self and my world", in that they have become heaviest, and even unbearable. And, from the ancient Prometheus to the contemporary Prometheus, technology continues to be our best ally. Understanding the reason why we feel "weaker" is easier and more reassuring when it is not a matter of "I" and "how I physically felt", which may be exceedingly complex, but nothing but a matter of "the apps" and "The feedback from that digital model". And, if it is true that we exponentially use "digital models" of our selves, from our health to our happiness and so forth, it is also true that what we may actually want to obtain is precisely to free ourselves from our core identity, specifically our autonomy, by externalising it from ourselves to technology—what we may actually want to obtain is precisely to trade our (heavier) core identity for a (lighter) digital identity: again, our (heavier) autonomy for technology's (lighter) automation.

An externalised, i.e. technological in the sense described above, self is lighter especially epistemologically and ethically. As we have seen in the case of the quantified self, a kind of externalisation from (internal) thought's uncertainty to (external) knowledge's certainty is pursued. More precisely, the move from the question on what our feelings are to the question on what our numbers are is most frequent. From an epistemological perspective, the advantage reported by self-trackers is that we do not stop trying to know our feelings, but we start actually knowing them by translating them into the numbers displayed by our self-tracking technologies: we make "guesswork", "gut", "inner worlds" and "vagaries of intuition" move from words to the numbers displayed by our self-tracking technologies. Finally, we obtain our "exosel[ves]" as our series of numbers.

And our "exosel[ves]" are far lighter than our selves for at least two reasons. First, if it is true that we believe that we can translate whatever into numbers, which are finally clearly displayed, it is also true that we believe that numbers can solve the issue of knowledge by translating complexity into easiness, specifically words' complexity, and even the complexity of what cannot be translated into words, into numbers' easiness. Second, if it is true that we believe that numbers' easiness can solve the issue of knowledge, specifically knowledge's complexity, it is also true that we believe that, finally, our decisions and actions can be equally easy: even though "There is no certainty", but "a mere probability", finding "the courage I need for my action" is easier if we believe that it can be based on (external) numbers easily readable, and not on (internal) mysteries difficultly readable. Moreover, finding "the courage I need for my action" is far easier if we also believe that there is a sense in which we are not desperately alone: the (external) numbers easily readable are external for a further reason, which is crucial—what is external can not only literally externalise the burdens of our decisions and actions from ourselves to something else but also give us something more than an ally: a scapegoat. Technology as what can shape our "exosel[ves]" is more than an ally through which we are not desperately alone when we should find "the courage I need for my action"—technology as what can shape our "exosel[ves]" is a scapegoat through which, from when we should find "the courage I need for my action" to when we should face its (irreducibly uncertain) outcomes, we unburden ourselves of our heaviest burdens, from our sense of responsibility, in the best case scenario, to our sense of guilt, in the worst case scenario.

And, again, sense of responsibility and sense of guilt strictly correlate with autonomy, as Kant masterfully teaches us through the correlation between autonomy, on the one hand, and being "culpable and deserving of punishment", on the other hand, which also correlates with "freedom". We may imagine a paradoxical scenario in which Reichenbach's Hamlet uses technology to decide what to do. We may imagine, for instance, that he uses a kind of truth algorithm. He inputs a series of data, from "good evidence" to "indirect evidence" and from the logician's "arguments" to the ghost's "arguments". He reads the output and acts accordingly. Whatever the output is, and, especially, whatever his action is, his question "But what if I should start thinking after the deed and find out I should not have done it?" becomes lighter to bear and easier to answer: Hamlet can say to both himself and others that he acted accordingly to what the truth algorithm displayed, which is far easier than saying to both himself and others that he acted accordingly to what he himself decided to do. Hamlet can prevent himself from being desperately alone not only when he should find "the courage I need for my action" but also when he should face its (irreducibly uncertain) outcomes—whatever his action is, Hamlet can say to both himself and others that it is not his fault, but the truth algorithm's. And, in a cultural context in which numbers are thought of as far more reliable than words, "guesswork", "gut", "inner worlds" and "vagaries of intuition" when it comes to obtaining knowledge, Hamlet's attempt to unburden himself of his heaviest burdens, from his sense of responsibility to his sense of guilt, is likely to succeed.

But what else happens to Hamlet? The key point is that unburdening himself of his heaviest burdens, from his sense of responsibility to his sense of guilt, means unburdening himself of his autonomy—externalising his sense of responsibility and his sense of guilt means externalising precisely the law (νόμος) Hamlet gives himself (αὐτός) as what causes his sense of responsibility and his sense of guilt: the more his action is caused by a heteronomous law, i.e. the truth algorithm's, the more he trades his autonomy's operating process for the truth algorithm's operating process. And the truth algorithm's operating process is nothing but automation—Hamlet trades his autonomy for what literally means, as we have seen, to "act of oneself, act offhand or unadvisedly", "to be done spontaneously or at random", "haphazard", to "introduce the agency of chance", "of things, [to] happen of themselves, casually", "to be self-produced" and, "of natural agencies, [to] act spontaneously".

The opposition between autonomy and automation is clearer if we think of Hamlet's two decision-making processes. In the case of autonomy, the ruling law is internalised, which means, first, that it is especially a matter of human capacities, starting with human rationality, and, second, that being alone is inescapable as something that the decision-making process requires (even though Hamlet asks the logician and the ghost, he is alone in the process of balancing their "arguments", "good evidence", "indirect evidence" and, especially, his "doubt". It is no coincidence that, not only in Shakespeare's case but also in Reichenbach's case, Hamlet's words coincide with a monologue, which is the most celebrated in the history of Western literature). Conversely, in the case of automation, the ruling law is externalised, which means, first, that it is especially a matter of technological capacities, starting with technological computation, and, second, that being alone is escapable: even though the action is Hamlet's, it is thought of as the output of the truth algorithm's decision-making process, in that, first, and quite explicitly, technological computation is thought of as better than human rationality from an epistemological perspective, in that it processes more data more quickly (which means that the former should replace the latter), and, second, and quite implicitly, technological computation is thought of as better than human rationality from an ethical perspective, in that sharing, and even escaping, individual responsibility and guilt is better than being alone (which means that, again, the former should replace the latter). Thus, even though the action is Hamlet's, it is not autonomous, but automated—even though the action is Hamlet's, it is the effect of a heteronomous law, which causes Hamlet's action and, consequently, makes it nothing but the last ring of a chain of cause and effect: the heteronomous law makes Hamlet's action nothing but the automated last ring of an automated chain of cause and effect.

Our technological "exosel[ves]" are frequently read as continuations of our selves.[21] My working hypothesis is that, day after day, sometimes quite explicitly and sometimes quite implicitly, we somehow want our technological "exosel[ves]" not to be the continuation of our selves, but the solution of continuity with our selves. Other phenomena (which I shall not address in what follows, but in future works) may be further promising cases in point, from the notion of avatar (whose etymology is, most interestingly, theological[22]) to the notion of digital twin. If my working hypothesis makes sense, then my argument according to which our technological era may be read as an overturning of human autonomy may be further developed as follows. On the one hand, technology, starting with the way we literally describe it (as we have seen especially in Chap. 2), obtains both autonomy as a key human prerogative and omnipresence, omniscience, omnipotence and inscrutability as key divine prerogatives. On the other hand, humans, starting with the way we literally describe them (for instance, if we think of the countless times in which we say that it is not up to us: it is up to the system, and our action is nothing but its effect), obtain automation as a key technological prerogative. Yet, even though the distance between technology and humans increases (in addition to my working hypothesis according to which, day after day, sometimes quite explicitly and sometimes quite implicitly, we want the former to be the solution of continuity with the latter), technology is

---

[21] References are several and interdisciplinary, starting with the notion of extended mind (see at least Clark & Chalmers, 1998).

[22] Oxford Dictionary of Phrase and Fable, see https://www.oxfordreference.com/view/10.1093/oi/authority.20110803095436845 (accessed in November 2021): "In Hindu belief, a manifestation of a deity or released soul in bodily form on earth; an incarnate divine teacher. The word comes from Sanskrit *avatāra* 'descent'".

still a humans' creation and humans are still technology's creators. Thus, most interestingly, something analogous to technology's meaning, as it is told starting with the cradle of Western culture, seems to emerge: the dimension of the creation, i.e. created, increasingly occupies the dimension of the creator. As far as the ancient Prometheus is concerned, Zeus is "dethroned" by "a son mightier than his father". As far as the modern Prometheus is concerned, Frankenstein, together with other humans, is indirectly one of the creature's victims. What may we say about the contemporary Prometheus? According to what we have seen in Chaps. 2 and 3, we may answer that, again, the contemporary Prometheus is increasingly replaced by the technologies he creates—more precisely, we may answer that the contemporary Prometheus wants the technologies he creates to increasingly replace him.

Finally, technology seems to increasingly occupy whatever ontological dimension, from human ontology to divine ontology, in addition to technological ontology. My own reading of the phenomenon described above is a kind of creators' surrender: since their burdens have become heaviest, and even unbearable, they create their creation, specifically technology, as their best ally, specifically scapegoat, to unburden themselves of them. But at least another reading may be added, which is not necessarily alternative to my own reading: a kind of creators' success precisely by surrendering to their creation, in the sense that creators may prove their power precisely by creating an even more powerful creation, to which to finally surrender. Thus, paradoxically enough, surrendering to one's own creation may be one's own power's, specifically power of creation's, best proof.

The infinite complexity of both humans and our technological era may be at least partially illuminated by challenging the equally infinite complexity of the following matching, which may seem quite paradoxical, and in which creation plays a key role. On the one hand, humans seem to overturn their being creations (which not necessarily implies a theological reading: humans as other humans' creations is sufficient) by creating something extraordinarily powerful, i.e. extraordinarily powerful technologies. On the other hand, humans seem to overturn the hierarchy between themselves and their extraordinarily powerful technologies by making the latter somehow create the former. Thus, our technological "exosel[ves]" may be read in two ways, which are not necessarily alternative. On the one hand, they are powerful to the point that they somehow create our selves, from being self-trackers, as we have seen, to being cyborgs. On the other hand, they are still creations, specifically our creations—more precisely, our technological "exosel[ves]" are still immanent, i.e. our artefactual creations, which means not only that anarchism may be a promising way to read them but also that we may continue to exploit them, even though they somehow create our selves, precisely by making them powerful to the point that they can be our best ally, and even our best scapegoat.

If the matching described above makes sense, then power, specifically power of creation, seems to play a key role in our technological era. History teaches us, first, that evolution is precisely a matter of power of creation and, second, that power of creation necessarily means power of destruction, in that what exists replaces what existed by evolving and what will exist will replace what exists by evolving. Thus, our urgent Hamletic question is the following: in the case of our own core identity's evolution, what do we imagine that we will want "To be or not to be"?

Part II
Challenges of the Future

Chapter 4
Epistemological Challenges

4.1 Philosophical Questions

The kinds of technologies upon which we have reflected, together with countless others, result from what we have defined as the epistemological move from the domain of *je ne sais quoi* to the domain of *logos* as "computation, reckoning". In addition to the testimonies of the quantified self upon which we have reflected, especially Carmichael's ("One pound heavier this morning? / You're fat. / 2 g too much fat ingested? / You're out of control. / Skipped a day of running? / You're lazy") and Williams' ("The feedback from that digital model often took precedence over how I physically felt. When I didn't eat 'enough' protein I felt weaker, and when I had too much sugar I felt fatter"), at least a further testimony entitled *28 Years of Tracking, But What Have I Learned?*[1] is worth quoting: "So I have learned that I'm not in fact an engineering problem of calories in and calories out. There is a lot more complex and subtle interactions going on that keep me constantly adjusting. What worked the first time didn't work the second time". The three testimonies show the idea according to which the radicalisation of *logos* as the only way to obtain reliable knowledge is questionable. In the first case, complexity, especially the meaning of being "out of control" and "lazy", is reduced to "2 g too much fat ingested" and "Skipped a day of running"—which means reducing complexity to simplism, worsened by the epistemologically anarchic absence of what we have defined as the expert's capacity to correctly subsume the particular to the universal. In the second case, complexity, especially the meaning of feeling "weaker", is reduced to "I didn't eat 'enough' protein"—which equally means reducing complexity to an analogous kind of simplism. In the third case, complexity, especially the meaning of "I'm", is reduced to "an engineering problem of calories in and calories out"—which equally

[1] See, also for the following quote, https://quantifiedself.com/show-and-tell/?project=815 (accessed in November 2021).

S. Chiodo, *Technology and the Overturning of Human Autonomy*,
Studies in Applied Philosophy, Epistemology and Rational Ethics 66,
https://doi.org/10.1007/978-3-031-26159-6_4

means reducing complexity to an analogous kind of simplism. If even regular self-trackers who believe in "self-knowledge through numbers" end up questioning the radicalisation of *logos* as the only way to obtain reliable knowledge, then what should we do with "computation, reckoning"?

The history of Western epistemology, from philosophical theories to scientific practices, coincides with the increasing success of *logos* as the best form of rationality to obtain reliable knowledge.[2] From ancient epistemology to modern epistemology to contemporary epistemology, *logos* has been winning over other possible ways to obtain reliable knowledge, from forms of irrationality to other forms of rationality. The philosophers we have considered as illuminating on the notion of autonomy, from Kant to Berlin, also stress rationality, especially rationality as *logos*. According to the former, "Every determining judgement is logical because its predicate is a given objective concept" (Kant, 1790: 20: 223) and, according to the latter, as we have seen, "I am rational, and [...] it is my reason that distinguishes me as a human being from the rest of the world". In addition to Kant and Berlin, it is worth quoting at least the greatest epistemological cornerstones: Plato ("the most perfect knowledge arises from the addition of rational explanation [*logos*] to true opinion", Plat, *Thaet.* 206 c), Aristotle ("we declare that the function of man is a certain form of life, and define that form of life as the exercise of the soul's faculties and activities in association with rational principle", Arist. *Eth. Nic.* I, 12, 1098a), Galilei ("I say that the human intellect does understand some of them [proposition] perfectly, and thus in these it has as much absolute certainty as Nature itself has. Of such are the mathematical sciences alone; that is, geometry and arithmetic", Gal. *Dialog.* I), Descartes ("I observed that this truth, I think, therefore I am (*cogito ergo sum*), was so certain and of such evidence that no ground of doubt, however extravagant, could be alleged", Descartes, 1637: IV), Cassirer on Einstein ("the general theory of relativity stands methodologically at the end of this series, since it collects all particular systematic principles into the unity of a supreme postulate, in the postulate not of the constancy of things, but of the invariance of certain magnitudes and laws with regard to all transformations of the system of reference", Cassirer, 1923: 404) and Wittgenstein ("3.03. Thought can never be of anything illogical, since, if it were, we should have to think illogically. 3.031. It used to be said that God could create anything except what would be contrary to the laws of logic. The truth is that we could not say what an 'illogical' world would look like", Wittgenstein, 1921: 3.03–3.031). As the series of quotes also show, the more we move from ancient epistemology to contemporary epistemology, the more two correlated phenomena emerge: first, the success of *logos* increases and, second, as a consequence of its increasing success, the more *logos* coincides with its strictest version, i.e. "computation, reckoning", the better it is thought to be at obtaining reliable knowledge.

Not surprisingly, the historical trajectory briefly described above results not only in the quantified self but also in countless current technologies based on "comput[ing], reckoning" algorithms: *logos* wins over whatever way to obtain reliable knowledge, from forms of irrationality to other forms of rationality. Yet, especially in the case of

[2] References are several, from Porter (1995) to Hansson (2018) to Schafer (2018).

correlations resulting from current technologies based on "comput[ing], reckoning" algorithms, *logos* has its Achilles' heel, as we have seen: "But shall I be right in this case?", which means that, whatever my decisions and actions are, they are necessarily nothing but "this case". And, even though "if something is probable I am allowed to make a posit and act as though it were true", in that "In doing so I shall be right in the greater number of cases", what I desperately want to know is "How then can I act in this case?", in that "I am afraid of doing something on the basis of a mere probability", which "has no meaning for an individual case". Moreover, what if *logos*, especially its strictest version, lacks something essential that cannot be "comput[ed], reckon[ed]"?

We should ask ourselves at least one last philosophical question: if we increasingly externalise *logos* from ourselves to technologies, in that they process more data more quickly (and, especially, in that they bear our epistemological and ethical burdens), then what is the destiny of both our *logos* in particular and our rationality in general? As we have seen, the way we use several current technologies shows that a key reason why we externalise *logos* from ourselves to technologies is that we increasingly consider the epistemological externalisation as the move from (internal) thought's uncertainty to (external) knowledge's certainty. The move described above questions the destiny of both our *logos* in particular and our rationality in general for at least two reasons. First, the more we deprive ourselves of circumstances in which we exercise our rational capacities, the more we risk atrophying them. Second, the more complex the circumstances we face are (for instance, globalisation and, consequently, global pandemics), the more uncertain our possible solutions are: again, the circumstances in which "There is no certainty", even though we use most sophisticated "predictive algorithm[s]", are countless, and "all I can reach" is "probability".

In what follows, I shall reflect upon plausible scenarios that may address both the issue of risking atrophying our rational capacities, which strictly correlate with our autonomy, and the issue of facing uncertainty,[3] which *logos* alone, both internalised and externalised, cannot solve.

[3] Current philosophical work on the notion of uncertainty is quite sophisticated. For the purpose of my monograph, it is sufficient to define uncertainty as follows: "By 'uncertain' knowledge, let me explain, I do not mean merely to distinguish what is known for certain from what is only probable. The game of roulette is not subject, in this sense, to uncertainty; nor is the prospect of a Victory bond being drawn. Or, again, the expectation of life is only slightly uncertain. Even the weather is only moderately uncertain. The sense in which I am using the term is that in which the prospect of a European war is uncertain, or the price of copper and the rate of interest twenty years hence, or the obsolescence of a new invention, or the position of private wealth-owners in the social system in 1970. About these matters there is no scientific basis on which to form any calculable probability whatever. We simply do not know" (Keynes, 1979: 113–114. See also Hansson, 1996).

4.2 Plausible Scenarios

Even the cradle of Western culture offers promising insights that can help us face uncertainty and, at the same time, not atrophy our rational capacities, which strictly correlate with our autonomy. Actually, *logos* is not alone even in the cradle of Western culture, in that it is joined at least by another form of rationality: *metis* (μῆτις), which literally means "wisdom, skill, craft, [...] counsel, plan, undertaking",[4] and which is joined by a further form of rationality, i.e. *phronesis* (φρόνησις), which literally means "purpose, intention, [...] practical wisdom, prudence in government and affairs".[5]

Century after century, *logos* has won over *metis*. But *metis* has not totally disappeared: it has been developed not only by ancient philosophers but also by modern and contemporary philosophers. As far as ancient philosophy is concerned, we can find *metis* in Plato's following words: if it is true that "what I do not know I do not think I know either" (Plat. *Apol.* 21 d), "For I was conscious that I knew practically nothing" (Plat. *Apol.* 22 d), it is also true that *metis*, i.e. wisdom, means "recogniz[ing] that he is in truth of no account in respect to wisdom" (Plat. *Apol.* 23 b). Thus, wisdom has to do with self-awareness of lack of knowledge (also, as far as *phronesis* is concerned, see Plat. *Prot.* 352 c and *Sym.* 209 a). Aristotle adds something that has to do with a kind of practical knowledge as the capacity to distinguish right from wrong when we have to act for the best: "men like Anaxagoras and Thales [...] while admitting them to possess a knowledge that is rare, marvellous, difficult and even superhuman, they yet declare this knowledge to be useless, because these sages do not seek to know the things that are good for human beings" (Arist. *Eth. Nic.* 1141 b. Also, as far as *phronesis* is concerned, see Arist. *Eth. Nic.* VI, 5). Stoic philosophers, especially Seneca, also work on wisdom's practical dimension: "as certain cliffs, projecting into the deep, break the force of the sea, and, though lashed for countless ages, show no traces of its wrath, just so the spirit of the wise man is impregnable, and has gathered such a measure of strength as to be no less safe from injury" (Sen. *Const.* III 5). Thus, wisdom has to do with being resilient to challenging obstacles. As far as modern philosophy is concerned, we can find wisdom in Descartes following words, which also stress the correlation between wisdom and a kind of practice-oriented knowledge: "the word 'philosophy' means the study of wisdom, and by 'wisdom' is meant not only prudence in our everyday affairs but also a perfect knowledge of all things that mankind is capable of knowing, for both the conduct of life and the preservation of health and the discovery of all manner of skills" (Descartes AT IXB 2), in addition to "its [wisdom's] teaching us to be masters of our passions and to control them with such skill that the evils which they cause are quite bearable, and even become a source of joy" (Descartes AT XI 488). Moreover, wisdom helps us understand "what decision it [the will] ought to make in each of life's contingencies"

[4] Liddell-Scott-Jones Greek-English Lexicon, see https://artflsrv03.uchicago.edu/philologic4/LSJ/navigate/47/1/1517/ (accessed in November 2021).

[5] Liddell-Scott-Jones Greek-English Lexicon, see https://artflsrv03.uchicago.edu/philologic4/LSJ/navigate/81/1/315/?byte=418670 (accessed in November 2021).

(Descartes AT X 361). Again, wisdom is a kind of practice-oriented knowledge that helps us master critical circumstances, from our decisions to our actions, even by making a virtue of necessity. As far as contemporary philosophy is concerned, after Fichte's stress on the relationship between wisdom and practice (see Fichte, 1794–1795), we can find wisdom in Nozick's impressive words: "What a wise person needs to know and understand constitutes a varied list: the most important goals and values of life—the ultimate goal, if there is one; what means will reach these goals without too great a cost; what kinds of dangers threaten the achieving of these goals; how to recognize and avoid or minimize these dangers; what different types of human beings are like in their actions and motives (as this presents dangers or opportunities); what is not possible or feasible to achieve (or avoid); how to tell what is appropriate when; knowing when certain goals are sufficiently achieved; what limitations are unavoidable and how to accept them; how to improve oneself and one's relationships with others or society; knowing what the true and unapparent value of various things is; when to take a long-term view; knowing the variety and obduracy of facts, institutions, and human nature; understanding what one's real motives are; how to cope and deal with the major tragedies and dilemmas of life, and with the major good things too" (Nozick, 1989: 269). Thus, again, wisdom has to do with a kind of practice-oriented knowledge that focuses on what we should do not only according to what circumstances we should know, from ourselves as single individuals to far more extended contexts of our decisions and actions, but also according to what values we should have. More precisely, three key points are stressed. First, being wise means giving priority to acknowledging and mastering disadvantageous circumstances, i.e. "dangers", "limitations", "tragedies" and "dilemmas", over acknowledging and mastering advantageous circumstances, i.e. "opportunities". Second, being wise means giving priority to adapting ourselves to circumstances, i.e. "obduracy of facts, institutions, and human nature", over adapting circumstances to ourselves. Third, being wise means giving priority to a kind of synthetic vision of the whole, i.e. "a long-term view", over a kind of analytic vision of the parts. Finally, other philosophers, starting with Kekes, stress the relationship between wisdom and a kind of practice-oriented knowledge: "To understand wisdom, we have to understand its connection with knowledge, action, and judgement" (Kekes, 1983: 277), in that, "in reliable, sound, reasonable, in a word, good judgement [...], a person brings his knowledge to bear on his actions" (Kekes, 1983: 277). Thus, wisdom is "The evaluative attitude [...] [that] is personal, not theoretical; anthropocentric, not metaphysical; context-dependent, not universal; and humanistic, not scientific" (Kekes, 2020.[6] See also Kekes, 1983; Maxwell, 1984; Zagzebski, 1996; Ryan, 1999; Tiberius, 2008; Vallor, 2016, 2017. And see also Ess, 2016; Gerdes, 2016; Sullins, 2016).

From ancient philosophy to contemporary philosophy, wisdom continues to be essentially defined as follows: first, self-awareness (especially of internal lacks and obstacles), second, awareness (especially of external lacks and obstacles), third, resilience, which also means being adaptable and making a virtue of necessity, and,

[6] Words reported as officially describing the monograph.

fourth, and especially, practice-oriented knowledge, which also means addressing the whole, from values to far more extended contexts.

If we go back to our issues, i.e. risking atrophying our rational capacities, which strictly correlate with our autonomy, and facing uncertainty that *logos* alone cannot solve, then we can find something promising in wisdom's essential definition. First, wisdom is a rational capacity to exercise. Second, wisdom as self-awareness and awareness of lacks and obstacles, resilience and, especially, practice-oriented knowledge can master domains that *logos* cannot master—which means that wisdom does not replace *logos*: conversely, the former is the latter's promising complementary form of rationality.

For instance, we may imagine to use wisdom as follows, from the case of Hamlet to the case of the dramatic pandemic we face. In the first case of uncertainty, i.e. Hamlet's, in which he has to decide on avenging his father, we may reflect as follows. First, wisdom as self-awareness (especially of internal lacks and obstacles) may recommend not relying on a ghost's words, in that Hamlet is shocked by his father's death. Second, wisdom as awareness (especially of external lacks and obstacles) may recommend not relying on a ghost's words, in that ghosts' existence is questionable. Third, wisdom as resilience, which also means being adaptable and making a virtue of necessity, may recommend exercising promising virtues to face uncertain circumstances, from carefulness against carelessness to patience against impatience. Fourth, and especially, wisdom as practice-oriented knowledge, which also means addressing the whole, from values to far more extended contexts, may recommend imagining long-term consequences (for instance, if the alleged murderer of Hamlet's father is guilty, then what may be the consequences of Hamlet's revenge on him, Hamlet, Hamlet's mother and Denmark? And, if the alleged murderer of Hamlet's father is innocent, then what may be the consequences of Hamlet's revenge on him, Hamlet, Hamlet's mother and Denmark?). Altogether, I think that wisdom may recommend not avenging Hamlet's father, as Shakespeare's wisdom itself masterfully teaches us.

In the second case of uncertainty, i.e. ours, in which we have to decide on being vaccinated as soon as the vaccine, whose long-term consequences are not known, is at our disposal, we may reflect as follows. First, wisdom as self-awareness (especially of internal lacks and obstacles) may recommend being mindful of our contingent fears, which can irrationally make us react in two far different ways to a novel vaccine (which we may reject) and to a novel medicine (which we may accept), even though, in both cases, long-term consequences are not known. Second, wisdom as awareness (especially of external lacks and obstacles) may recommend not giving priority to what we know less, i.e. the long-term consequences of the vaccine, over what we know more, i.e. the dramatic consequences, both short-term and long-term, of the pandemic on several issues, both individual and social, from health to education to economy to fairness. Third, wisdom as resilience, which also means being adaptable and making a virtue of necessity, may recommend exercising, both as experts, i.e. scientists and politicians, and as citizens, promising virtues to face uncertain circumstances, from farsightedness against shortsightedness to humility against narcissism (which also means, in the case of experts, renouncing media exposure if it is confusing). Fourth, and especially, wisdom as practice-oriented knowledge, which also means addressing the whole, from values to far more extended contexts, may recommend imagining long-term consequences (for instance, if the vaccine will turn out not to be sufficiently safe in the long run, and we would have decided to be vaccinated, in my case, and not to be vaccinated, in your case, then what may be the consequences of our actions on our health, morality, society and future generations, if we consider that we would have accepted, in my case, and rejected, in your case, to test on ourselves something that may be helpful to others in the future, which is precisely what others did for us in the past? And, if the vaccine will turn out to be sufficiently safe in the long run, and we would have decided to be vaccinated, in my case, and not to be vaccinated, in your case, then what may be the consequences of our actions on our health, morality, society and future generations, if we consider that we would have accepted, in my case, and rejected, in your case, to test on ourselves something that may be helpful to others in the future, which is precisely what others did for us in the past?). Altogether, I think that wisdom may recommend to be vaccinated as soon as the vaccine, whose long-term consequences are not known, is at our disposal.

What is most important to note is that, in both cases, i.e. Hamlet's and ours, "There is no certainty" at all, and, yet, it seems that wisdom can do what *logos* cannot do, which is most essential: "giving me the courage I need for my action"—the most essential reason why wisdom should be added to *logos* as its promising complementary form of rationality is that it can make us act even though we have "no certainty" at all, in that it can give us reasons to act that can be valid even though they cannot be "true", "certain". And, in our era's complexity and uncertainty, having valid reasons to act even though we do not have "true", "certain" reasons to act can even save us, from our health to our morality and from ourselves as single individuals to our society. More precisely, from the perspective of *logos* as "computation", which is what technology can offer (at least for the present), reasons to act are valid if they are at least probabilistically "true", "certain". Conversely, from the perspective of *metis* as "wisdom", which is what humans can offer (at least for the present), reasons to act are valid if there are self-awareness and awareness of lacks and obstacles, resilience and, especially, practice-oriented knowledge—the more our era is characterised by complexity and uncertainty, the more we may happen to be helped by *metis* as "wisdom" even more than by *logos* as "computation", in that the former's domain addresses complexity and, especially, uncertainty even more than the latter's domain (for instance, what would we do with correlations, especially if they are countless, resulting from even more countless data? We may happen to need great wisdom to decide and act among countless correlations).

A further most important thing is worth noting in both cases, i.e. Hamlet's and ours: imagination is continuously implied (see at least Kind & Kung, 2016; Kind, 2018; Badura & Kind, 2021). In the case of wisdom as self-awareness, imagination is required in terms of being capable of identifying not only actual internal lacks and obstacles but also potential internal lacks and obstacles. In the case of wisdom as awareness, imagination is required in terms of being capable of identifying not only actual external lacks and obstacles but also potential external lacks and obstacles. In the case of wisdom as resilience, imagination is required in terms of being capable of identifying the way in which it is possible both to be adaptable and to make a virtue of necessity. Moreover, in the case of wisdom as practice-oriented knowledge, imagination is required in terms of being capable of identifying the whole, from values to far more extended contexts, especially in terms of long-term consequences. Thus, we may add something more to wisdom's essential definition we have obtained from the history of Western philosophy: since imagination is continuously implied, we may define wisdom as the capacity to imagine a series of things, i.e. possible lacks and obstacles, possible ways both to be adaptable and to make a virtue of necessity and, especially, possible identifications of the whole, from values to far more extended contexts, especially in terms of long-term consequences—we may define wisdom as the capacity to imagine especially what we cannot know as at least probabilistically "true", "certain", i.e. what is uncertain, according to Keynes (as we have seen), in that "About these matters there is no scientific basis on which to form any calculable probability whatever. We simply do not know".

If my proposed definition of wisdom makes sense, then its difference from *logos* becomes clearer. We may metaphorically say that, if it is true that wisdom walks on quicksand, it is also true that it can walk precisely where *logos* cannot walk—and quicksand is not only most extended but also increasing in our unprecedentedly complex and uncertain era. We may add that wisdom is more horizontal and *logos* is more vertical: the latter can make us see actual details, but the former can make us imagine potential scenarios of relationships between details and far more extended contexts. Thus, we may add that wisdom is more farsighted and *logos* is more shortsighted: the latter's speciality is making us see the present, but the former's speciality is making us imagine the future. Finally, we may add that wisdom is more altruistic and *logos* is more egoistical: the latter, being more vertical and shortsighted, can make us see especially our own present interests, to which we end up giving priority over everything else, but the former, being more horizontal and farsighted, can make us imagine especially our and others' future interests, to which we end up giving priority over everything else.

Interestingly enough, as we have seen by quoting Plato ("civic wisdom he had not, since this was in the possession of Zeus"), even according to the cradle of Western culture, humans give priority to what is singular and individual, i.e. details they can see by using *logos*, over what is plural and common, i.e. scenarios of relationships between details and far more extended contexts they can imagine by using wisdom. But, as globalisation's complexity in general and pandemic's uncertainty in particular show us, what our era demands more than ever is precisely Plato's "civic wisdom" as the capacity to address what exceeds whatever is singular and individual—which means, from an epistemological perspective, the capacity to exceed what we know is there, and especially to move to what we imagine will be there (again, the capacity to imagine especially what we cannot know as at least probabilistically "true", "certain"), and, from an ethical perspective, the capacity to exceed what matters to us, and especially to move to what will matter to others.

Actually, wisdom is a form of rationality that, being especially practice-oriented knowledge, but also resilience, combines an epistemological attitude and an ethical attitude. Wisdom as the capacity to address the whole, from values to far more extended contexts, means not only the epistemological capacity to imagine long-term consequences as a series of logical steps but also the ethical capacity to imagine long-term consequences as a series of potential circumstances demanding ethical resources of us (on the complexity of ethical deciding and acting, see, for instance, McDermott, 2008). Ethics' complexity exceeds logic's complexity in the sense that it is a matter of deciding and acting not only by reflecting upon what is true and what is false but also, and especially, by reflecting upon the infinitely complex relationship between what is true and what is false, on the one hand, and what is right and what is wrong, on the other hand (and right and wrong extend from the individual dimension to the social dimension). When Hamlet questions himself on his action's potential consequences on himself, the alleged murdered of his father, his mother and Denmark, he uses not only logical steps but also ethical resources, from ethical evaluations (for instance, is it right to avenge? When is it right? And when is it wrong? Moreover, "But shall I be right in this case?") to empathy (for instance, what will happen to Hamlet, the alleged murdered of his father, his mother and Denmark? And how will they feel?). And, when we question ourselves on our actions' potential consequences on our health, morality, society and future generations, we use not only logical steps but also ethical resources, from ethical evaluations (for instance, is it right to test on ourselves something that may be helpful to others in the future? When is it right? And when is it wrong? Moreover, "But shall I be right in this case?") to empathy (for instance, what will happen to our health, morality, society and future generations? And how will they feel?).

If it is true that wisdom demands ethical resources of us, from ethical evaluations to empathy, it is also true that it is a form of rationality that can be even more complementary to *logos*, in that it can give us reasons to act that can be even ethically valid even though they cannot be "true", "certain". We may go back to my real example, i.e. the letter from the hospital in which it is literally written that, according to the "predictive algorithm" that processed the data from the questionnaire you filled in, you are not eligible to take the specific genetic test, in that your risk of being genetically predisposed to suffer from the specific cancer is not high enough. What if we react by using *logos* alone? And what if we react by using both *logos* and wisdom? In the first case, we are likely to behave by following Hamlet's logician's instructions, according to which "if something is probable I am allowed to make a posit and act as though it were true". Speaking of a risk that is not high enough means speaking of a risk that is not "probable" enough. Thus, "I am allowed to make a posit", according to which the most "probable" thing is not to be genetically predisposed to suffer from the specific cancer, "and act as though it were true", which may discourage from preventing severe health problems that may happen. In the second case, we are likely to behave by following what results from integrating Hamlet's logician's instructions with something else, which is precisely the capacity to imagine possible lacks and obstacles, possible ways both to be adaptable and to make a virtue of necessity and, especially, possible identifications of the whole, from values to far more extended contexts, especially in terms of long-term consequences. For instance, we may identify as a possible internal lack our attitude according to which, year after year, we end up taking an absence of risk instead of a risk that is not "probable" enough. Thus, wisdom may help us have valid reasons to continue our prevention. And we may identify as a possible external lack public healthcare's attitude according to which, if something is not "probable" enough, then it ends up moving from public healthcare to private healthcare. Thus, wisdom may help us have valid reasons not to take an absence of risk instead of a risk that is not "probable" enough to be taken over public healthcare. More precisely, wisdom may help us have valid reasons to continue our prevention by possibly using private healthcare at least from time to time. And so forth.

My real example is analogous to several other real examples, starting with the pandemic. One of the key reasons why "A severely outdated pandemic plan could have contributed to thousands of Covid-19 deaths in Italy" (Giuffrida & Boseley, no page number[7]) and "Britain found itself at the back of the global queue for essential pandemic-fighting equipment, including masks and other protective equipment for health staff and carers and vital chemicals for tests" (Giuffrida & Boseley, no page number), may be explained by using Hamlet's logician's words: "if something is probable I am allowed to make a posit and act as though it were true". Again, *logos* alone may make us fall into dramatic unpreparedness, in that what is not "probable" enough may make us "act as though it were" false, which means what follows: "Coping with a pandemic without good plans, with insufficient intensive care skills and a scarce stock of health protection devices, is like finding yourself driving a bus on a mountain road in the middle of a sudden and heavy snowfall without snow chains" (Giuffrida & Boseley, no page number). We may say that, conversely, integrating *logos* with wisdom means having "snow chains" even though the risk of "finding yourself driving a bus on a mountain road in the middle of a sudden and heavy snowfall" is not "probable" enough. In the case of the pandemic, wisdom as the capacity to imagine possible lacks and obstacles, possible ways both to be adaptable and to make a virtue of necessity and, especially, possible identifications of the whole, from values to far more extended contexts, especially in terms of long-term consequences, may help us as follows. First, as far as imagining possible lacks and obstacles is concerned, wisdom may lead us to think of globalisation as something that can unprecedentedly change reality as we know it. Thus, we may work on preparedness more than ever, from imagining unprecedented scenarios (in which phenomena characterising other geographical areas may quickly characterise ours) to prepare basic resources, starting with training (which means both training at least selected individuals to face unprecedented scenarios and preparing plans). Second, as far as imagining possible ways both to be adaptable and to make a virtue of necessity is concerned, wisdom may lead us to think of unprecedented scenarios as opportunities to change and improve what is normally most difficult to change and improve (for instance, the pandemic seems to encourage us to reconsider cornerstones of our societies, from our relationships with nature to our economies). In addition, wisdom may lead us to train our adaptability, together with our imagination itself, by making us imagine our possible reactions to unprecedented scenarios

[7] See, also for the following quotes, https://www.theguardian.com/world/2020/aug/13/italy-pandemic-plan-was-old-and-inadequate-covid-report-finds (accessed in November 2021).

(which is also exceedingly advantageous in our normal life, even though we will not face other unprecedented scenarios). Third, and especially, as far as imagining possible identifications of the whole, from values to far more extended contexts, especially in terms of long-term consequences, is concerned, wisdom may lead us to think of unprecedentedly complex and uncertain circumstances, starting with the pandemic, as opportunities to reconsider ourselves both as single individuals and as a society. For instance, if we think of our actions as something that can quickly impact, more than ever, several others' lives synchronically, i.e. not only in our geographical areas but also in others', then what actions of ours should we reconsider as right? And what actions of ours should we reconsider as wrong? And, if we think of our actions as something that can quickly impact, more than ever, several others' lives diachronically, i.e. not only in the present but also in the future, then what actions of ours should we reconsider as right? And what actions of ours should we reconsider as wrong? More precisely, in both cases, what values should we choose as leading our actions? For instance, if we think of the global pandemic, which is both severely complex and severely uncertain, then what values (from legitimate kinds of egoism to legitimate kinds of altruism) should we choose as leading our actions (from rejecting vaccines whose long-term consequences are not known, even though others tested on themselves vaccines that can save our lives in the present, to accepting vaccines whose long-term consequences are not known, not only in that others tested on themselves vaccines that can save our lives in the present but also in that we may save others' lives in the future)?

The questions can further stress that exercising wisdom can help us not only face uncertainty but also not atrophy our rational capacities, which strictly correlate with our autonomy: trying to find answers surely help us not deprive ourselves of circumstances in which we exercise our rational capacities, which strictly correlate with our autonomy, especially if we add wisdom, which also implies imagination, to *logos*.

We should add one last key question, which more directly makes us go back to technology: can technology be wise (see Dalal et al., 2016 on a correlated issue)? As we have seen, there is a sense in which technology can be even extremely logical, but is there a sense in which it can be wise? At least currently, the answer seems negative, if we define wisdom as the capacity to imagine possible lacks and obstacles, possible ways both to be adaptable and to make a virtue of necessity and, especially, possible identifications of the whole, from values to far more extended contexts, especially in terms of long-term consequences—at least currently, technology does not seem capable of being wise, if we define wisdom as the capacity to imagine especially what we cannot know as at least probabilistically "true", "certain", i.e. what is uncertain. More precisely, technology seems to lack the core of wisdom's essential definition, which, from its history (as self-awareness and awareness of lacks and obstacles, resilience and, especially, practice-oriented knowledge) to my proposed definition (as the capacity to imagine what is uncertain), is a matter of imagination—the key reason why technology does not seem capable of being wise is that is seems to lack imagination, from both an epistemological perspective and an ethical perspective.

Imagination is a complex philosophical notion, but, for the purpose of our reflection upon our relationship with technology, we may at least say that technology seems to lack what scholars of imagination define, first, as the kind of capacity to understand other minds that also implies empathy (see at least Nichols & Stich, 2003; Goldman, 2006; Carruthers, 2009. On the relationship between imagination and empathy see at least Heal, 2003; Stueber, 2011), second, as the capacity to think in a creative way (see at least Gaut, 2003; Stokes, 2016; Kind, 2022), third, as the capacity to obtain knowledge through thought experiments (see at least Kind & Kung, 2016; Kind, 2018) and, fourth, as the capacity to understand metaphorical language (see at least Walton, 1990). We may go back to my real example of the "predictive algorithm". What the "predictive algorithm" can do is to correlate our data with a great number of data. If the correlations between our data and critical data are not high enough, then we are not eligible to take the specific genetic test, in that our risk of being genetically predisposed to suffer from the specific cancer is not high enough. The "predictive algorithm" is surely powerful: it makes us aware of correlations of which, otherwise, we would not be aware, and which can give us crucial clues to better understand diseases' causes and promising therapies. Yet, again, the Hamletic question demands something more of us: if "a probability has no meaning for an individual case", i.e. ours, then shall we "be right in this case?". The "predictive algorithm" and Hamlet's logician share the kind of answer they can give us: "No answer", in the sense that, even though we "will be right in the greater number of cases", we shall not "know it for certain". Again, we face uncertainty. And, again, wisdom may help us, as we have seen, by giving us valid reasons to continue our prevention by possibly using private healthcare at least from time to time.

Wisdom can help us even more definitely if we face circumstances that are even more severely uncertain, from exceptions to novelties. On the one hand, exceptions have to do with what does not fall under the domain of "the greater number of cases", according to Hamlet's logician. For instance, we may imagine that our case is the exception if, even though it correlates with "the greater number of cases" found by the "predictive algorithm", it does not fall under their domain: the domain of "the greater number of cases" coincides with past cases that turned out to be negative (no disease emerged), but our case turns out to be positive (the disease emerges), even though it has several correlations with "the greater number of cases". Exceptions' key points are at least two. First, when we happen to be exceptions, we stop thinking of them as successful technologies' collateral damage. Conversely, we start thinking of them as successful technologies' lacks to solve—and wisdom may be promising to solve successful technologies' lacks. Second, even though we do not happen to be exceptions, the more global and complex our society is, the greater the number of exceptions we face as a society is. In Kant's Königsberg, the number of exceptions to face was actually limited to what happened in Königsberg and in Europe. What happened in other geographical areas impacted minimally and slowly. In our society, almost whatever happens (and almost wherever it happens) impacts us, even maximally and quickly: the number of exceptions to face is potentially unlimited for us—what does not fall under the domain of "the greater number of cases" found by the "predictive algorithm" is potentially unlimited for us, which means that, again, wisdom can complementarily work with *logos*, in that the former can promisingly lead us to face exceptions the latter cannot promisingly lead us to face. On the other hand, novelties have to do with what "We simply do not know", according to Keynes, in that "About these matters there is no scientific basis on which to form any calculable probability whatever". We do not even need to imagine what novelties may be: our era is strongly characterised by even disruptive novelties, from the pandemic to climate change to the key issue of my monograph, i.e. the sense in which current technologies can overturn human autonomy in particular and human identity and dignity in general. If it is true that "predictive algorithm[s]" "are based on the past, and on the assumption that patterns will repeat" (O'Neil, 2016: 38), it is also true that we cannot be prepared to face novelties if we totally move from using a series of forms of rationality, from *logos* to wisdom, to using *logos* alone, specifically its externalisation from the human mind to technology. The reason why "predictive algorithm[s]" alone cannot prepare us to face novelties is not only that they "are based on the past" but also that, if we totally move decision-making processes from the human mind to technology, then we deprive decision-making processes themselves of being promisingly affected, sometimes consciously and sometimes unconsciously, by the human mind's capacities we scarcely know, from forms of rationality won by *logos*, starting with wisdom, to capacities that do not fall under the domain of rationality. For instance, we may happen to make our best decision after having slept on it precisely because we let the human mind's capacities we scarcely know promisingly affect our decision-making processes—but, if we totally externalise our decision-making processes from the human mind to technology, then we deprive ourselves of the possibility of exploiting whatever does not fall under the domain of *logos*, specifically the form of *logos* that can be externalised from the human mind to technology.

The aim of the second part of my monograph is asking philosophical questions, which is one of the classic tasks of philosophy, even more than answering. Yet, the plausible scenarios described above start answering. If the philosophical questions are on the epistemological challenges of the future, then I believe that complexity and uncertainty are our most crucial epistemological challenges of the future, and that *logos* alone, specifically its externalisation from ourselves to technology, cannot make us prepared. More precisely, I believe that *logos* alone, specifically its externalisation from ourselves to technology, can be successful whenever the balance between at least probabilistic certainty and uncertainty is acceptable, which means, for instance, that the mistakes we may make can be remediable (if an algorithm tells us to buy a book, and we understand that the book is not right for us after having read ten pages, then we can resell the book). But I believe that *logos* alone, specifically its externalisation from ourselves to technology, cannot be successful whenever the balance between at least probabilistic certainty and uncertainty is not acceptable, which means, for instance, that the mistakes we may make cannot be remediable. If an algorithm tells us that we are not eligible to take a specific genetic test, then we should have the guarantee of a dialogue with a doctor who can possibly revise the algorithm's output—starting with using wisdom. And, again, if an algorithm tells us that we are not eligible to have a given job, a given insurance and a given mortgage, then we should have the guarantee of a dialogue with a given staff, from the human resources' to the insurance company's to the bank's, who can possibly revise the algorithm's output—starting with using wisdom, which can easily acknowledge bias, differently from *logos* externalised from ourselves to technology, and, especially, which can easily promote fairness against discrimination.[8]

[8] References are almost countless (see at least O'Neil, 2016 for an articulate series of cases in point).

Chapter 5
Human Identity's Challenges

5.1 Philosophical Questions

Epistemological challenges lead us to reflect upon human identity's challenges, in that the former imply something that is thought of as underpinning the latter: the exercise of rational capacities that make us not only potentially autonomous but also, and especially, actually autonomous. As we have seen, from ancient philosophy to contemporary philosophy, especially in the last three centuries, autonomy is one of the most essential cornerstones of the definition itself of human identity in Western culture. From the ancient idea of a kind of rational self-determination to the modern idea of autonomy especially as the capacity to give oneself laws to the contemporary idea of autonomy especially as a kind of individualism, autonomy is a privileged way not only to define what humans are, i.e. capable of self-giving laws (which is an idea strongly characterising both modern philosophy and contemporary philosophy), but also to strictly correlate their being capable of self-giving laws with both their morality in general and their moral actions in particular, i.e. "commend[ing] justice", "work[ing] out the good" and "ha[ving] a character". Thus, what may be the destiny of human identity if we increasingly move autonomy, as one of the most essential cornerstones of our definition itself as humans, from ourselves to technology?

Trading humans' autonomy for technology's automation does not mean at all that technology can be autonomous in the same way as humans can be autonomous, as it is specified even in the introduction. Technology as we know it cannot be autonomous in the ways in which the history of Western philosophy thinks of autonomy: from rational self-determination to the capacity to give oneself laws to individualism, technology lacks what is most essential, i.e. its own reasons—technology as we know it can have no reasons but humans' reasons. Whenever machine learning algorithms result in surprising outputs, the reason why their outputs are surprising is not a matter of machine learning algorithms' own reasons. For instance, we cannot say that, even though we expected a negative output, the output was positive because of the machine learning algorithm's optimism. And so forth. Technology as we know it

S. Chiodo, *Technology and the Overturning of Human Autonomy*,
Studies in Applied Philosophy, Epistemology and Rational Ethics 66,
https://doi.org/10.1007/978-3-031-26159-6_5

always represents humans' reasons—technology as we know it always lacks humans' autonomy as their special kind of causing (technological) effects and outputs, in that humans' special kind of causing (technological) effects and outputs is always a matter of making sense of human life.

But there is a sense in which technology can be autonomous, which is the key issue of my monograph: technology can be autonomous in the ways we think of it, specifically in the ways we define it (as we have seen, autonomous systems, autonomous software, autonomous devices, autonomous applications, autonomous machines, autonomous vehicles, autonomous equipment, autonomous drones, autonomous weapons, autonomous robots, autonomous agents, autonomous workloads and so forth). Consequently, technology can be autonomous in the ways we use it (as we have seen especially in Chaps. 2 and 3)—which means what is most essential for the purpose of our reflection upon our relationship with technology: technology can be autonomous in the sense that it can be thought of, defined and used by us as our perfect way to escape from our own autonomy by making technology be our perfect scapegoat, i.e. our own autonomy's perfect avatar (whose etymology is, most interestingly, theological, as we have seen).

And the consequence of our making technology be our own autonomy's perfect avatar is that, even though technology cannot be actually autonomous, i.e. autonomous in the same way as humans can be autonomous, we can actually lose our own autonomy—we can actually become more and more automated, i.e. technology's automation's perfect avatars.

The following objection may be raised: if there is only a partial sense in which technology can be autonomous, then there is only a partial sense in which we can be automated precisely because, differently from technology, we have our own reasons as our special kind of causing effects that is always a matter of making sense of human life. But what if we lose our own reasons as our special kind of causing effects that is always a matter of making sense of human life?

5.2 Plausible Scenarios

In Chap. 3, we have imagined a paradoxical scenario in which Reichenbach's Hamlet uses a truth algorithm to decide what to do: first, he inputs a series of data, from "good evidence" to "indirect evidence" and from the logician's "arguments" to the ghost's "arguments", and, second, he reads the output and acts accordingly. We have reflected upon the fact that Hamlet obtains a series of advantages: he is not desperately alone when he should act and face his action's outcomes and, especially, he can say to both himself and others that it is not his fault, but the truth algorithm's—which means that he obtains the most essential advantage, i.e. to unburden himself of his heaviest burdens, from his sense of responsibility to his sense of guilt. But there is a further advantage Hamlet obtains: he can be more "efficient", which literally means "Making, causing to be; that makes (a thing) to be what it is; chiefly in connection with cause. [...] Productive of effects; effective; adequately operative".[1] Hamlet can be more "efficient" not only by "Making, causing to be", in that he can quickly move from doubting to acting, but also by being "Productive of effects" and "adequately operative", in that he can do other things a prince should do, after having quickly moved from doubting to acting. The truth algorithm makes Hamlet capable of doing more things more quickly. We may also imagine that, thing after thing, Hamlet starts giving priority to doing more things more quickly, in addition to the other advantages described above, over doubting itself as a way to shape his own reasons—Hamlet starts giving priority to efficiency over doubting itself as a way to make sense of his life.

If we move from Hamlet to ourselves, then actual and potential scenarios in which we give priority to efficiency over doubting itself as a way to make sense of our lives exponentially increase. From housewives to CEOs, we give priority to doing more things more quickly (from housewives' daily activities to CEOs' daily activities) over doubting itself as a way to shape our own reasons (from housewives' thoughts on domestic issues to CEOs' thoughts on international issues)—and the sense in which we can actually become more and more automated, i.e. technology's automation's perfect avatars, is precisely the following: we exponentially trade our doubts, as ways to shape our own reasons that make sense of our lives, for technological efficiency, as a way to do more things more quickly.

Examples in which we can find the phenomenon described above are several. They can help us better understand the sense in which we can actually become more and more automated, which is not only what happens whenever we google something, in that we frequently rely on what Google's algorithm makes appear on the first page, but also what happens whenever the quantified self's self-trackers as nonexperts use their self-tracking technologies. In the first case, the correlation between efficiency and automation is clear. Whenever we rely on what Google's algorithm makes appear on the first page, one of our key reasons is being "Productive of effects" and "adequately operative", i.e. efficient: even though we know that what Google's algorithm makes appear on the first page does not result from our own present intentions, but from others' past intentions, we rely on it because of its advantages, from experiencing a kind of reassuring sharing to doing more things more quickly. And making others' past intentions replace our own present intentions through technology's automation means automating ourselves, in that what we actually do is to use technology to make others' reasons replace our own reasons that make sense of our lives—what we actually do is to use technology to move from autonomy as the capacity to give ourselves laws to automation as effects of causes "offhand", "unadvised", "spontaneous", "random", "haphazard", "introduc[ing] the agency of chance", "casual", "accident[al]" and "without any regard to the merit of man", as we have seen (see Chap. 2). In the second case, the correlation between efficiency and automation is even clearer. Whenever the quantified self's self-trackers as nonexperts use their self-tracking technologies, one of their key reasons is being

[1] Oxford English Dictionary, see https://www.oed.com/view/Entry/59742?redirectedFrom=efficient#eid (accessed in November 2021).

"Productive of effects" and "adequately operative", i.e. efficient (as their testimonies frequently report, see Chap. 3). Moreover, even though they seem effects of their own causes, their nonexpert solipsism makes them fall into being effects of causes "offhand", "unadvised", "spontaneous", "random", "haphazard", "introduc[ing] the agency of chance", "casual", "accident[al]" and "without any regard to the merit of man". If we, as nonexpert self-trackers, use our smartphones to count our daily steps according to the belief that they should be 10.000 a day, but we do not consider the contexts that experts, i.e. doctors, would consider (not only our general health conditions but also, and especially, their expertise in reading the relationship between our general health conditions and taking 10.000 steps a day), then we risk harming our health by automating our relationship with it—then we risk harming our health by making our relationship with it the automated effect of causes that are, again, "offhand", "unadvised", "spontaneous", "random", "haphazard", "introduc[ing] the agency of chance", "casual", "accident[al]" and "without any regard to the merit of man", in that they result not from an expert thoughtfulness, but from a nonexpert randomness as the effect of causes that are not, and cannot be, our own: we are the contexts to consider, but the numbers displayed by our smartphones are out of context in the sense described above.

Clearly enough, speaking of becoming more and more automated strictly correlates with speaking of becoming more and more anarchic: even from an etymological perspective, what we lose by moving from being autonomous to being automated is precisely what can actually rule, i.e. what the etymological root of autonomy includes as νόμος and what the etymological root of anarchism excludes as άρχω. But what may it mean to become totally automated (and anarchic)? As we have seen, we even seem to want automation and anarchism to be our escapes from our own autonomy's unbearable epistemological and ethical burdens by making technology be our perfect scapegoat, i.e. our own autonomy's perfect avatar. Thus, may we imagine a scenario in which to become totally automated (and anarchic) means to improve the quality of human life, and possibly to evolve as humans?

Something from the past may help us answer the question: the notion of "autarky" (αὐτάρκεια, transliterated as *autarkeia*), which shares the first half of its etymology, i.e. αὐτός as "itself", with both the notion of autonomy and the notion of automation, but which introduces something meaningful through the second half of its etymology, i.e. ἀρκέω (transliterated as *arkeo*) as "to be strong enough, to be sufficient, to suffice",[2] finally meaning "self-sufficiency".[3] What is meaningful is that "autarky" as "self-sufficiency" is precisely the core of the wise's life in several ancient philosophies: Cynical, Cyrenaic, Epicurean, Stoic and Sceptical. Even though "autarky" as "self-sufficiency" is differently characterised by different philosophies, something essential is shared: "autarky" as "self-sufficiency" strictly correlates with the wise's

[2] Liddell-Scott-Jones Greek-English Lexicon, see http://www.perseus.tufts.edu/hopper/morph?l=%E1%BC%80%CF%81%CE%BA%CE%AD%CF%89&la=greek#lexicon (accessed in November 2021).

[3] Liddell-Scott-Jones Greek-English Lexicon, see http://stephanus.tlg.uci.edu/lsj/#eid=17980 (accessed in November 2021).

life, which is finally happy. For instance, Seneca writes: "I'll give you a ready-made backer; in the famous phrase of Cato, 'borrow from yourself'. No matter how small the loan, it'll be enough if we seek from ourselves whatever we lack. Lucilius, my friend, it makes no difference whether you feel no need of something or you have it already. In either case the upshot is the same: you will not be in anguish. Nor do I instruct you to deny something to nature—she is unyielding, she is unbeatable, she demands her due—but rather I instruct you to be aware that whatever goes beyond nature is at the whim of others and not necessary" (Sen. *Ep.* 119, 2). We may translate Seneca's proposal into what follows: if we face complexity and uncertainty, which can easily make us "be in anguish", then "borrow[ing] from yourself" and "seek[ing] from ourselves whatever we lack" is wiser than relying on something external, whatever it may be, in that what is external is out of our control. It is no coincidence that the wise's happy life results from something internal: "Decide that you are happy when all of your joy comes from within you, when you gaze upon the things which people seize, wish for, protect and yet find nothing which you would—I don't say 'prefer', but nothing you would want. I'll give you a brief guideline by which you can measure yourself, by which you can tell that you have become complete: you will only have what is yours when you come to understand that the least fortunate are fortunate" (Sen. *Ep.* 124, 24). Thus, if it is true that the wise's happy life results "from within you", i.e. "becom[ing] complete" through "only hav[ing] what is yours", it is also true that it results from self-sufficiency—the wise's happy life results from autarky.

Yet, even though humans seem to continue to pursue happiness century after century, they do not equally seem to continue to pursue autarky century after century. Conversely, the historical moves from the ancient Prometheus to the modern Prometheus to the contemporary Prometheus show that humans increasingly externalise their prerogatives and capacities from themselves to technology—more precisely, what humans increasingly externalise from themselves to technology is autarky's most advanced outcome, i.e. autonomy.

Autarky and autonomy do not coincide at all, but there is a meaningful sense in which the former anticipates the latter. From an etymological perspective, when "to be strong enough, to be sufficient, to suffice" can be said about ourselves (which is autarky's etymological definition), we can start giving ourselves laws (which is autonomy's etymological definition). It is no coincidence that the less "strong", i.e. autarchic as self-sufficient, we are, the more we risk falling into heteronomy as autonomy's antonym. For instance, we may risk falling into heteronomy when our critical thinking is not "strong enough", but quite weak, which makes us end up being ruled by someone else. From a philosophical perspective, when the imperatives "borrow from yourself", "seek from ourselves whatever we lack" and "become complete" through "only hav[ing] what is yours" can be said about ourselves (which is Seneca's philosophical definition of autarky), we can start not only exercising "Autonomy of the will" as "the property of the will by which it is a law to itself" (which is Kant's philosophical definition of autonomy) but also acting "as if the maxim of your action were to become by your will a universal law of nature" (which is Kant's philosophical definition of categorical imperative). For instance, we are autonomous in making decisions and acting if the reasons underpinning our making decisions and acting are "borrow[ed] from" ourselves. Conversely, we are heteronomous in making decisions and acting if the reasons underpinning our making decisions and acting are others' reasons—which may also mean that we are heteronomous in making decisions and acting if the reasons underpinning our making decisions and acting are technology's outputs.

Actually, technology can severely challenge our autarky and autonomy, starting with the following basic attitude, which is quite typical of technology designers: whatever issue may be there, it is an issue to solve (through technology). Thus, we end up having at our disposal whatever technological solution to whatever issue—more precisely, we end up having at our disposal whatever technological solution to what philosophers, differently from technology designers, may consider as opportunities to leave there precisely as they are, and not as issues to solve (through technology). A real example may be the following: I happened to talk to the owner of a technology company about the pleasure of walking in the mountains, which he shared with me, and he ended up saying that walking in the mountains would have been even more pleasant if we could have used automated mountain trails that could have deprived us of the effort to walk. My real example is clearly paradoxical, but its logic is shared by several other examples. For instance, we may think of cases in which we use technological solutions to deprive ourselves of whatever kind of physical effort (which seems to make sense). But, after having deprived ourselves of whatever kind of physical effort (for instance, through technological solutions depriving ourselves of whatever kind of housework, and even of opportunities to walk on foot), we say that our health needs physical effort. Thus, paradoxically enough, we do not go back to housework, and even to opportunities to walk on foot, which we continue to shift from ourselves to technological solutions, but we pay for going to the gym (which seems to make less sense), where we use other technological solutions depriving ourselves of the most challenging kinds of physical effort (which seems to make even less sense. See Hansen, 2020 on a correlated issue). And, especially, we may think of cases in which we use technological solutions to deprive ourselves of whatever kind of mental effort. Examples are several, from architects, i.e. creatives, who use technological solutions to design houses (softwares can allow architects to input a series of data, and, finally, to obtain an architectural design almost ready to use[4]), to writers, i.e. creatives, who use technological solutions to write articles (softwares can allow writers to input a series of data, and, finally, to obtain an article almost ready

[4] See, for instance, https://academy.archistar.ai/top-ten-design-software-for-architects (accessed in November 2021), which reports the list of "the top ten design software for architects": "design software automates common processes. This allows you to save time. You can get designs to clients quicker and enjoy a more efficient workflow. [...] Its key feature is automation. [...] the software offers several tools for automating complicated processes. For example, it has an inbuilt library of stair and rail designs. [...] You only need to enter a number into your variable to create the desired number of elements". See also Carr (2014).

to use[5]). Even more interestingly, we have at our disposal technological solutions to design houses if we are not architects, i.e. if we are nonexpert ("Undertaking a major renovation or new construction without an architect to provide design know-how or detailed plans would seem to be the realm of either the foolish or the very handy. [...] But in recent years, a spate of computer software programs has made it possible for homeowners to skip the architect altogether and do the design work themselves", Kurutz, 2012, no page number), and technological solutions to write articles if we are not writers, i.e. if are nonexpert (advertising of writing softwares for beginners is frequent, especially stressing that writing softwares are easy and quick[6])—again, current technologies are frequently clear symptoms of epistemological anarchism.

Moreover, current technologies are frequently clear symptoms of the shift from humans to technology of what has been defined for millennia as something essentially human not only in terms of prerogatives and capacities but also, and especially, in terms of what (pleasantly) makes sense of human life. The key reason why architects and writers as creatives may decide not to use technological solutions to deprive themselves of creativity's mental effort, from designing houses to writing articles, is not that designing houses and writing articles are essentially human prerogatives and capacities (as we have seen, several softwares can somehow design houses and write article)—conversely, the key reason why architects and writers as creatives may decide not to use technological solutions to deprive themselves of creativity's mental effort is that, as human history shows us, being creative can (pleasantly) make sense of human life: being creative is even one of the best ways humans have at their disposal to (pleasantly) make sense of human life.

The issue raised by the relationship between technological solutions and human creativity is key, in that it can make us understand something crucial: until humans will constitutively need to make sense of their own life, there will be something unshiftable from humans to technology—until making sense of human life will constitutively be humans' key issue, there will be something unshiftable from humans to technology.

Yet, from technological solutions that replace human creativity to the kinds of technologies upon which we have reflected as replacements of human autonomy,

[5] See, for instance, https://hostingtribunal.com/best/article-generator-software/#gref (accessed in November 2021), which reports the list of "the best article generator software": "WordAI understands your source material – whether it's in English, Spanish, French, or Italian – and rephrases it as a human would. It's a smart, accurate, and easy-to-use content generator, great for spinning scrapped articles or generating fresh articles for SEO. [...] SEO Content Machine is a full-featured app that helps you generate, illustrate, and publish high-ranking content. Built with SEO in mind, this AI also provides link building tools and impressive integrations with popular software. [...] Just give Articoolo a topic, and it will generate a unique article for you. The AI gathers sources online and structures a piece for any niche – all at the click of a button. Plus, Articoolo can also help you spin old content for an SEO boost or summarize articles to pack value for readers".

[6] See at least https://www.blogbrandz.com/tips/writing-software/ (accessed in November 2021) and https://xtendedview.com/internet/apps-for-beginner-writer/5147/ (accessed in November 2021).

we seem to increasingly undermine, sometimes consciously and sometimes unconsciously, human making sense of human life—more precisely, we seem to increasingly undermine human autarky as the condition of possibility for human autonomy as the condition of possibility for human making sense of human life.

Actually, we may read the examples of architects and writers as creatives as follows. First, by increasingly using softwares designing houses and writing articles, architects and writers increasingly become less autarchic as self-sufficient when it comes to being creative, in that creativity, as whatever capacity, requires exercise not only to evolve but also not to atrophy (for instance, I, as a writer of philosophical works, know from experience that the more I make the effort to shape philosophical ideas by my own writing, the more creative the philosophical ideas I can shape are, in that making the effort means exercising and, consequently, evolving and not atrophying my own creativity, from my capacity to imagine to my capacity to shape). Second, by increasingly becoming less autarchic as self-sufficient when it comes to being creative, architects and writers increasingly become less autonomous when it comes to deciding what they should do and why, from what kind of houses to design and for what reasons to what kind of articles to write and for what reasons, in that what they should do, specifically what they can do and why are heteronomously decided by their softwares (for instance, I, as a writer of philosophical works, know from experience that the more autarchic as self-sufficient I am when it comes to shaping philosophical ideas, the more autonomous I am when it comes to deciding what kind of philosophical ideas to shape and for what reasons, in that my range of possibilities is decided by the kind of philosophical ideas I myself can shape accordingly to my own reasons, and not by the kind of philosophical ideas my softwares can shape accordingly to their own processes). Thus, being autarchic as self-sufficient and, consequently, autonomous also means being professionals, and not amateurs, in that the kind of things done and the reasons why they are done are not (unaware) effects of causes coinciding with softwares' (heteronomous) decisions, but (aware) effects of causes coinciding with humans' (autonomous) decisions. Third, by increasingly becoming less autonomous when it comes to deciding what they should do and why, architects and writers increasingly become less capable of making sense of human life, i.e. their own lives, in that the kind of things they do and the reasons why they do them are increasingly divided from their own selves—we increasingly become less capable of making sense of human life, i.e. our own lives, if our autonomy, as our capacity to do the kind of things we ourselves can do accordingly to our own reasons, is increasingly divided from our own selves, i.e. our own identities as humans.

We may imagine a scenario in which our own selves, i.e. our own identities as humans, totally become automated, which means that they totally lose their autonomy as their capacity to do the kind of things they themselves can do accordingly to their own reasons. Current technologies surely help us imagine the scenario described above, starting with a kind of increasing division between our own selves as characterised by their being irreducibly analog (at least to a great extent) and our digital selves. For instance, when we sleep, the former, being irreducibly analog (at least to a great extent), are lying bodies and dreaming minds that are not aware of what the latter are at the same time: not lying and dreaming, but receiving both emails and likes and dislikes. We may say that, when our own selves as characterised by their being irreducibly analog (at least to a great extent) sleep, our digital selves do not sleep at all: conversely, our digital selves are totally awake, from receiving both emails and likes and dislikes to being seen and read to making money and making others make money. A further example may be clearer. In the last years, I have increasingly shifted my memory from myself to technology, sometimes by using analog technology, i.e. paper agendas, and sometimes by using digital technology, i.e. digital agendas. Several years ago, when I had to remember something, I was autarchic and autonomous especially in the sense that I myself could remember what I had to remember without depending on something else, from lost paper agendas to broken digital agendas. And, if it is true that humans can lose their memory, it is also true that they happen to lose their agendas more than their memory. Today, when I have to remember something, I am not autarchic and autonomous at all—conversely, I am automated especially in the sense that what I do is caused by what my agendas tell me to do. Thus, I am not autarchic and autonomous at all not only in the sense that, if my paper agendas get lost and my digital agendas get broken, then I cannot act but also in the sense that, if hackers sabotage what my digital agendas tell me to do, then my actions' reasons are not mine, but theirs. Moreover, there are meaningful senses in which I am not capable of making sense of human life, i.e. my own life. On the one hand, losing my capacity to remember means impoverishing my daily life in terms of having at my disposal less opportunities to awake images, emotions, insights, thoughts, arguments and decisions actually underpinning my own making sense of my own life. On the other hand, acting by being driven by others' reasons means losing the following essential opportunities to make sense of my own life: first, the opportunity to better understand who I am actually by experiencing my own actions as a kind of my own reasons' embodiments (for instance, I can better understand to what extent I am actually courageous when I move from imagining myself in dangerous circumstances to actually acting in dangerous circumstances) and, second, the opportunity to better understand and plan who I want to be and why, i.e. my own future self, my own future identity, which is one of the best ways to make sense of my own life (for instance, if I better understand and plan that I want to be more courageous and, especially, that the reasons why I want to be more courageous are a matter of both self-improvement and contributing to social improvement when it comes to fairness, then, whatever action I will do, it, as a kind of my own reasons' embodiment, will greatly make sense of my own life).

According to several authors (see at least Domingos, 2015), our increasing use of technological solutions to whatever kind of issue means almost exclusively advantages, starting with having more time at our disposal to nurture human relationships and spirituality. It is possible. But the critical issue I can discern is, again, a matter of risking increasingly atrophying our autarky, our autonomy and, finally, our capacity to make sense of our own lives. I believe that there is a meaningful difference between the following two scenarios. In the first scenario, whatever kind of issue is solved by technology. Thus, we are free to focus our energies exclusively on nurturing human relationships and spirituality without facing obstacles. The scenario described above is promising, but I believe that the second scenario is even more promising. In the second scenario, technology does not solve whatever kind of issue. For instance, we continue not to be deprived of the kinds of physical and mental efforts we have seen. On the one hand, we are not free to focus our energies exclusively on nurturing human relationships and spirituality without facing obstacles, precisely in that we have to face and possibly solve obstacles that sometimes are even most challenging. But, on the other hand, the more we have to face and possibly solve challenging obstacles, the more we have the opportunity to do something crucial: to exercise something that may be defined as a meta-capacity, being precisely the capacity to awake whatever capacity, from our physical capacities to our mental capacities, to face and possibly solve challenging obstacles that unpredictably happen to us—I believe that the key reason why the wisest thing we can do is not to shift whatever issue to solve from ourselves to technology is that, otherwise, we would lose our most crucial capacity, which is precisely the meta-capacity to face uncertainty (which, day after day, more strongly characterises our era), also by being capable of bearing efforts.

Thus, we should choose what to shift, even totally, from ourselves to technology and what not to shift, even at all, from ourselves to technology. Promising distinguishing criteria may be the following, for the reasons we have seen: we may choose not to (totally) shift from ourselves to technology, first, what can make us (even pleasantly) make sense of our own lives (which also requires what can make us exercise our autarky and autonomy) and, second, what can make us exercise the meta-capacity to face uncertainty (which also requires what can make us capable of bearing efforts).

And, if the objection is that, when we will be cyborgs (in the strictest sense of the term, i.e. as beings whose mechanical and digital dimensions far exceed the other dimensions), both making sense of our own lives and uncertainty will dissolve, then the answer is that humans as we know them, i.e. we ourselves, are not cyborgs, in the sense that, even though, as even the ancient Prometheus teaches us, humans are constitutively technological, their own selves are (still) characterised by their being irreducibly analog in the sense that we have seen. A further objection may be that we should do our best to stop being analog. But what is not totally clear to me is the reasons why stopping being analog would be surely better than being as we are. For instance, I cannot even imagine the reasons why using automated mountain trails that could deprive me of the effort to walk would be surely better (and more pleasant) than walking in the mountains. Analogously, I cannot even imagine the reasons why using softwares that could deprive me of the effort to write articles would be surely better (and more pleasant) than writing articles.

Thus, on the one hand, I think that we definitely have valid reasons to continue to exercise our capacities, i.e. our autarky, our autonomy, our capacity to make sense of our own lives and, especially, our meta-capacity to face uncertainty, which, day after day, more strongly characterises our era. On the other hand, I think that we do not definitely have valid reasons to stop exercising them, starting with the following logical paradox: our future as cyborgs (in the strictest sense of the term, i.e. as beings whose mechanical and digital dimensions far exceed the other dimensions) is possible, but not certain at all—thus, since we continue to face uncertainty again and again, we equally continue to definitely have valid reasons to exercise our autarky, our autonomy, our capacity to make sense of our own lives and, especially, our meta-capacity to face uncertainty itself by not totally shifting them from ourselves to technology.

In addition to the advantages described above, there is a further advantage in exercising our capacities, specifically our autonomy, which we should not totally overturn by shifting it from ourselves to technology in the sense that we have seen. As Kant masterfully teaches us, autonomy plays a key role in underpinning something that becomes even more crucial in our era of globalisation—autonomy plays a key role in underpinning not only human identity but also human dignity, freedom and morality (see Chap. 1), which globalisation continuously challenges. More precisely, being autonomous and, consequently, dignified, free and moral also means contributing positively to society, in that, as we have seen, "when we think of ourselves as free we transfer ourselves into the world of understanding as members of it and cognize autonomy of the will along with its consequence, morality". We may imagine two scenarios, the first one coinciding with Kant's Königsberg and the second one coinciding with our global society. In both cases, narcissistic intellectuals say that the virus that circulates is less severe than a flu. We may also imagine, in our thought experiment, that, in both cases, the virus that circulates is the SARS-CoV-2 virus (about which it was actually said that it was less severe than a flu[7]). The narcissistic intellectuals are doctors, but not virologists and, especially, not virologists expert in the SARS-CoV-2 virus (in the last two years, several doctors have actually talked about it even though they were not virologists expert in it, and they did not renounce personal media exposure even though it was confusing). The question we should ask ourselves is the following: in what scenario is autonomy more crucial? It is easy to answer that autonomy is more crucial in the second scenario, i.e. our global society, than in the first scenario, i.e. Kant's Königsberg, in that, even though intellectuals should always control their narcissism, they are responsible for a series of consequences that, moving from the first scenario to the second scenario, impact people both more extensively and more dramatically—the more global our society is, the more responsible intellectuals should be, which means that they should be more autonomous: autonomy, as the capacity to act accordingly to one's own reasons, is what can make one responsible for one's own actions. Moreover, in our global society, autonomy may be thought of as more crucial than ever not only when it is intellectuals' but also when it is people's. People's heteronomy is, again, both more extensive and more dramatic in the second scenario than in the first scenario. One of the most dystopian scenarios we may imagine coincides with a kind of global heteronomy. For instance, a global heteronomy meaning that billions of people are driven by an elite's reasons, which may coincide with an elite's interest—and, for instance, a global heteronomy meaning that billions of people are driven by technology's kinds of reasons, which may coincide with no one's interest.

[7] See, for instance, https://www.theguardian.com/science/2020/jan/27/china-coronavirus-who-to-hold-special-meeting-in-beijing-as-death-toll-jumps (accessed in April 2020), which reports an article entitled *Like the Flu? Trump's Coronavirus Messaging Confuses Public, Pandemic Researchers Say*, "The Guardian", March 13, 2020.

But is it possible to imagine a scenario in which we are driven by technology's kinds of reasons, i.e. no one's interest, after having lost our autonomy? I believe that there is at least a meaningful sense in which it is possible. Even though, as we have seen (and as the current debate on technology's nonneutrality stresses, see at least Noble, 2018), technology has no reasons but humans' reasons, the meaningful sense in which we are driven by technology's kinds of reasons, i.e. no one's interest, is a matter of the kind of randomness that automation and anarchism share—we are driven by technology's kinds of reasons, i.e. no one's reasons, whenever our actions result from something that stops coinciding with reasons and starts coinciding with causes, specifically technology's random causes, which happens whenever technology's outputs start being, meaningfully, autonomous, specifically autonomous from our inputs. For instance, whenever there is an inscrutable gap between coders' reasons and machine learning algorithms' outputs, there is also a meaningful sense in which we are driven by technology's kinds of reasons, i.e. no one's reasons, in that humans' reasons are replaced by technology's random causes, i.e. outputs—the more technology becomes, meaningfully, autonomous from humans' reasons, starting with coders' reasons, the more we are driven by nothing but technology's random outputs, which replace what can be defined as actual reasons: humans' reasons underpinning humans' making sense of their own lives.

Again, a kind of trade emerges: technology trades its automation for our autonomy (at least in the meaningful sense described above) and we trade our autonomy for technology's automation (at least in the meaningful sense described above). Yet, our era, characterised by unprecedented complexity and uncertainty, may be the worst time to lose our autonomy—thus, again, one of the most urgent philosophical questions of our era is the following: why do we trade precisely what may save us from our era's unprecedented complexity and uncertainty? And, especially, why do we trade it precisely for what may make us be overpowered by our era's unprecedented complexity and uncertainty?

We should definitely have valid reasons, starting with what upon which we have reflected: the irresistible rise of autonomy as humans' core identity becomes, sometimes consciously and sometimes unconsciously, an unbearable burden. Thus, on the one hand, we have countless individuals who try to unburden themselves of their unbearable burdens by trading their autonomy for technology's automation and, on the other hand, we have an elite of individuals who try to capitalise on it. But the countless individuals and the elite of individuals share something essential: they are equally users of technological solutions that, by increasingly solving whatever issue, increasingly make them exercise less their autarky, autonomy, capacity to make sense of their own lives and meta-capacity to face uncertainty. Actually, the pandemic seems to show that humans, from common people to entrepreneurs to politicians, have the following strengths and weaknesses. Their most important strength is, again, a matter of technological solutions, starting with surprisingly quick and effective vaccines. And their most important weakness is a matter of lack of capacities, starting with autarky, autonomy, capacity to make sense of their own lives and meta-capacity to face uncertainty as also underpinning what follows: a wise physical and mental balance that, even in the most difficult circumstances, can obtain from autarky the strength of self-sufficiency, from autonomy the strength of dignifying, free and moral actions done accordingly to their own reasons, from the capacity to make sense of their own lives the strength of authentic living and from the meta-capacity to face uncertainty the strength of courageous living—their most important weakness is a matter of lack of what in Brontë's masterpiece is represented as the capacity of the one "who stands so grave and quiet at the mouth of hell, looking collectedly at the gambols of a demon" (Brontë, 1847: 258).

Autarky, autonomy, capacity to make sense of our own lives and meta-capacity to face uncertainty can make a difference in addressing our era's complexity and uncertainty both as single individuals and as a society. As we shall see, autonomy, together with the capacities that strictly correlate with it, may also serve as a kind of antidote against individualism (in the best case scenario) and anarchism (in the worst case scenario) that worsen our era's most dramatic challenges.

Chapter 6
Ethical Challenges

6.1 Philosophical Questions

Even autarchy is thought of as something that may also serve as a kind of antidote against individualism. Actually, even though autarchy can lead to individualism according to the Cynics and the Cyrenaics, it can lead to something quite different according to the Stoics. Epictetus writes that "when he [Zeus] chooses to be the Giver of rain and the Giver of fruits, and the Father of Gods and men, you see that he cannot obtain these functions and these names, if he is not useful to man; and, universally, he has made the nature of the rational animal such that it cannot obtain any one of its own proper interests, if it does not contribute something to the common interest" (Epict. *Discourses* 1). And Marcus Aurelius writes that, since "man is formed by nature to acts of benevolence, when he has done anything benevolent or in any way conductive to the common interest, he has acted conformably to his constitution, and he gets what is his own" (M. Aur. *Med.* 9). Not only "the good for the reasonable animal is society" (M. Aur. *Med.* 5. Which means that "so long as nothing of the kind drives me out, I remain, am free, and no man shall hinder me from doing what I choose; and I choose to do what is according to the nature of the rational and social animal", M. Aur. *Med.* 5) but also "The intelligence of the universe is social" (M. Aur. *Med.* 5). Not surprisingly, moving from autarky to autonomy, the idea of something that may also serve as a kind of antidote against individualism develops. According to Kant, autonomy serves as the condition of possibility for the categorical imperative that, through its transcendental structure, finally serves as the condition of possibility for moral duties towards both oneself and others: "I say that the human being and in general every rational being exists as an end in itself, not merely as a means to be used by this or that will at its discretion; instead he must in all his actions, whether directed to himself or also to other rational beings, always be regarded at the same time as an end" (Kant, 1785: 4: 428), which "is an object of respect" (Kant, 1785: 4: 428). Thus, Kant's philosophical work on the notion of autonomy is also fundamental to address ethical challenges both as single individuals

S. Chiodo, *Technology and the Overturning of Human Autonomy*,
Studies in Applied Philosophy, Epistemology and Rational Ethics 66,
https://doi.org/10.1007/978-3-031-26159-6_6

and as a society, in that autonomy underpins the practical imperative: "So act that you use humanity, whether in your own person or in the person of any other, always at the same time as an end, never merely as a means" (Kant, 1785: 4: 429).

Yet, autonomy's irresistible rise especially from modern culture to contemporary culture does not go together with an equally irresistible rise of a kind of (respectful) altruism resulting from Kant's practical imperative. Conversely, even though Kant's practical imperative can serve as a kind of antidote against (disrespectful) individualism, Plato's words about humans seem to continue to win over considering themselves "never merely as a means": as we have seen, humans do not have "civic wisdom", which means that they seem constitutively incapable of giving what is plural and common priority over what is singular and individual. Paradoxically enough, contemporary philosophers even work on autonomy as a kind of individualism, especially procedural individualism, which means that autonomy moves from a transcendental structure underpinning a kind of intersubjectivity to a subjective structure underpinning a kind of individualism. For instance, according to coherentism (as what defines autonomy in terms of coherence between what actually determines one's action and how one actually judges it from an ethical perspective, as the following references show), autonomy is a matter of a kind of subjective coherence, in which subjective preferences play a key role: "A preference, as I shall use the term, is a very particular sort of desire: it is one for a certain first-level desire to be effective in action, when or if one acts, and that is formed in the search for what is good. A preference, that is, has as its intentional object the state of affairs of a certain of one's first-level desires being satisfied in action, and it is formed by an agent's evaluating that first-level desire with respect to some standard of goodness. This evaluation need not take place at the conscious level" (Ekstrom, 1993: 603. See also Bratman, 2007). Thus, on the one hand, autonomy can lead to what can be defined, together with Plato, as "civic wisdom", by stressing humans' focus on what they share, starting with autonomy's transcendental structure. On the other hand, autonomy can lead to what can be defined, together with contemporary philosophers, as a kind of individualism, by stressing humans' focus on what they do not share, starting with what is subjective. But, if subjectivity is stressed again and again, then it can lead to what finally overturns autonomy itself: it can lead to anarchism—the stress on subjectivity can lead to anarchism as what finally overturns autonomy itself, in that stressing subjectivity again and again can mean stressing a contingent, and even idiosyncratic, randomness into which whatever ruler finally dissolves.

In what follows, I shall reflect upon plausible scenarios that may address the issue of human autonomy's ethical challenges in our technological era. More precisely, my questions shall be the following: if our era, characterised by unprecedented complexity and uncertainty, may be the worst time to lose not only our autonomy in general but also our autonomy as underpinning our intersubjectivity in particular (precisely as what finally underpins our capacity to face global, i.e. shared, challenges), then what may be the philosophical meaning of the paradoxical overturning of our autonomy? And, especially, what may be, from an ethical perspective, the philosophical meaning of our paradoxical trade of our autonomy for technology's automation?

6.2 Plausible Scenarios

The current relationship between technology and health may serve as a promising case in point. Interestingly enough, our era is also described as the era of personalised medicine, which, for instance, the European Union defines as "a medical model using characterisation of individuals' phenotypes and genotypes (e.g. molecular profiling, medical imaging, lifestyle data) for tailoring the right therapeutic strategy for the right person at the right time, and/or to determine the predisposition to disease and/or to deliver timely and targeted prevention. Personalised medicine relates to the broader concept of patient-centred care, which takes into account that, in general, healthcare systems need to better respond to patient needs" ("Official Journal of the European Union" 2015[1]). Advantages are clear, starting with the repetition of the word "right": personalised medicine can "tailor" preventions and therapies to single individuals, i.e. "the right person at the right time" (it is no coincidence that the word "patient" is also repeated: "patient-centred care" and "patient needs").

Interestingly enough, we have introduced the notion of tailoring when we have talked about a kind of tailored-made knowledge in the case of self-tracking technologies quantifying health (see Chap. 3). If we move from an epistemological perspective to an ethical perspective, then we understand that the notion of tailoring is an even more meaningful symptom of our technological era—the notion of tailoring increasingly characterises our use of technology.

Even though the advantages of personalised, i.e. tailored-made, medicine are clear, we should question the philosophical, specifically ethical, meaning of its increasing use. As we have seen in the case of self-tracking technologies quantifying health, a kind of individualism is stressed. For instance, when Wolf and De Groot define their notion of personal science, they continuously repeat words implying a kind of individualism: "Personal science specifically addresses personal questions. We use the word personal in its ordinary language sense, pointing to questions directly relevant to the individual asking the questions, and concerning their private life, experiences, and emotions. It is the self-reflexive quality of the questions that makes personal science personal; that is, the researcher's own life is the research domain. Personal science always involves the deliberate choice of the individual about what questions to ask, what method to use and what observations to make" (Wolf & De Groot, 2020: 2). And, as we have seen, when the Quantified Self Institute describes itself, the words used are the following: the quantified self "is the gathering of personal data for and by you. […] It is a functionally 'selfish' activity, which is the result of a personal motivation. 'Me and my data', that is the point with Quantified Self". Finally, in addition to scholars according to whom, as we have seen, "the choice to actively track herself is, as she put it, 'liberating'", according to other scholars, self-tracking is not only symptomatic of a kind of individualistic epistemology ("personal science is a very self-related affair", Heyen, 2020: 133. And "the knowledge that […] [self-trackers] gain from their n = 1 experiments is highly particularistic and

[1] See https://eur-lex.europa.eu/legal-content/EN/TXT/?uri=OJ%3AC%3A2015%3A421%3AFULL (accessed in November 2021).

idiosyncratic", Sharon, 2017: 109) but also symptomatic of a kind of individualistic ethics ("appeal of self-tracking can only be understood when viewed against the modern narcissistic quest for uniqueness and exceptionalism", Morozov, 2013: 233[2]).

Self-tracking technologies in general and the quantified self in particular definitely differ from personalised medicine. Actually, the former are anarchic in a way that is extraneous to the latter, in that personalised medicine continues to use what self-tracking technologies and the quantified self do not continue to use, at least in most cases: experts' expertise (which self-tracking technologies remove whenever doctors are replaced by nonexperts). Yet, both self-tracking technologies and the quantified self, on the one hand, and personalised medicine, on the other hand, are characterised by a kind of individualism that is extraneous to past ways to address health.

Individualism emerges from both an epistemological perspective and an ethical perspective. As far as the epistemological reading is concerned, both phenomena are characterised by the risk of an increasing removal of comparisons. In the case of self-tracking technologies and the quantified self, as we have seen in Chaps. 3 and 4, the increasing removal of comparisons (as an actual risk) even leads to a kind of nonexperts' epistemological solipsism. In the case of personalised medicine, the idea is precisely that "tailoring the right therapeutic strategy for the right person at the right time" is incomparably better than using standard protocols. Again, advantages are clear and worth pursuing. But, according to wisdom's farsightedness, we should also address the following risk (as a potential risk): the idea described above may potentially lead to an increasing removal of comparisons—and increasingly removing comparisons means increasingly removing opportunities to find both actual mistakes and potential mistakes, i.e. even better ways to "tailor [...] the right therapeutic strategy for the right person at the right time". More precisely, increasingly removing comparisons means risking increasingly thinking of a single particular case as incommensurable with other cases: literally, a single particular case risks being increasingly thought of as not measurable against other cases (which is epistemologically meaningful in the era of the rise of *logos*). Doctors as experts' potential epistemological risk is ending up absolutising a particular case by underestimating both other particulars cases and, especially, the universal both as the whole and as a possible rule shared by whatever case (which means, again, ending up removing opportunities to find both actual mistakes and potential mistakes). People as nonexperts' potential epistemological risk is ending up absolutising their own particular cases and, consequently, falling into practical risks: for instance, the risk of overestimating their own particularities as potential symptoms of potential health problems (conversely, knowing something less about themselves may be wiser than knowing whatever particularity about themselves, in that their own particularities may be more easily de-dramatised by standard protocols and more easily dramatised by personalised medicine even though they are not meaningful) and the risk of underestimating

[2] See also Cressey (2017). Quite exhaustive references to do-it-yourself (DIY) science are in Ferretti (2019). In addition, Dickenson (2013) extends her reflection to potentially negative uses of technology in a more general context by analysing the relationship between what may be defined as "me medicine" against what may be defined as "we medicine".

public healthcare, in that private healthcare may easily (and purposely) seem better suited for personalised medicine.

As far as the ethical reading is concerned, the increasing removal of epistemological comparisons characterising both phenomena lead to an equally increasing removal of ethical comparisons. In the case of self-tracking technologies and the quantified self, as we have seen in Chaps. 3 and 4, the increasing removal of epistemological comparisons even leads to a kind of nonexperts' socio-existential solipsism. In the case of personalised medicine, the idea is precisely that, if it is true that "individuals' phenotypes and genotypes (e.g. molecular profiling, medical imaging, lifestyle data)" are incomparably better than ideal models resulting from abstracting from other individuals, it is also true that other individuals are less meaningful to individuals—it is also true that, when it comes to defining one of the most fundamental characteristics of humans' core identity, i.e. human health, the sense of what is singular and individual wins over the sense of what is plural and common, which means that being irreducibly single individuals wins over being a society in which something essential is shared and, consequently, strengthens the sense of community itself. But, again, our era, characterised by unprecedented complexity and uncertainty, may be the worst time to lose what can contribute to underpin our intersubjectivity in general and our sense of community in particular. The more incommensurable with other "individuals' phenotypes and genotypes" my "phenotypes and genotypes" are, the less I am interested in interacting with them: if there is almost nothing to share with them, then there is almost no reason to interact with them. When my "phenotypes and genotypes" bring me good news, I am not urged to share it with my relatives, in that I am likely to think that their "phenotypes and genotypes" make what they do not share with me win over what they share with me. And, when my "phenotypes and genotypes" bring me bad news, I am not urged to share it with my friends, in that I am likely to think that I have almost nothing to share with them: again, I have almost no reason to interact with them, which also means that both my opportunities to passively experience empathy (as the object of others' empathy for me) and my opportunities to actively experience empathy (as the subject of my empathy for others) decrease—and decreasing my opportunities to experience empathy means increasingly stopping exercising it and, consequently, increasingly starting atrophying it. In addition, socio-existential solipsism stops the kinds of opportunities to evolve that are urged by comparisons with others. For instance, if I, as a hypochondriac, risk dramatising my own particular conditions, then comparisons with others, who are not hypochondriac, can help me learn not to dramatise only if I consider their particular conditions as worth comparing with mine, in that we have something essential to share. Conversely, if I consider their particular conditions as extraneous to mine, then I deprive myself of opportunities, first, to empathically experience that there are ways not to dramatise, second, to learn not to dramatise and, third, to evolve not only as a hypochondriac in particular (who increasingly learn to be more wisely balanced) but also as a human being in general (who increasingly learn what may be the wisest thing to learn: again, the meta-capacity to face uncertainty as the capacity of the one "who stands so grave and quiet at the mouth of hell, looking collectedly at the gambols of a demon"). Finally, there is a sense in which socio-existential solipsism undermines society itself, especially global society itself. The more I think of myself as irreducibly unique, the less I think of others as possible representatives of my own irreducible uniqueness in several contexts: not only healthcare but also whatever context is addressed by politics, from rights, starting with education, to duties, starting with respect.

Thus, even though personal medicine's advantages are worth pursuing, philosophers' wisest contribution to doctors may be the following idea: the more we pursue personal medicine's advantages, the more we should nurture ways in which we can exercise our intersubjectivity in general and our sense of community in particular. For instance, we may continue to ask our doctors both to also use diagnostic comparisons (as opportunities to find actual and potential mistakes) and to also talk to us in person (in that epistemological and socio-existential solipsism may lead to a further bureaucratisation of doctors' role: when what matters is data as most sophisticated technologies' outputs, doctors risk not having reasons to talk to us in person, as my real example of the "predictive algorithm" shows).

Moreover, we should nurture ways in which we can exercise our intersubjectivity in general and our sense of community in particular for the following key reason: the more we think of ourselves as irreducibly unique, the less we think of others as possible representatives of our own irreducible uniqueness in a specific context, i.e. representative democracy—the more we absolutise individuals' uniqueness without balancing it with individuals' commonality, the more representative democracy risks being undermined and even dissolving into political anarchism (and our notion of human health, as one of the most fundamental characteristics of humans' core identity, may play a key role). The logical steps of representative democracy's risk of being undermined and even dissolving into political anarchism are the following. The first logical step of representative democracy's risk is that, if we think of ourselves as irreducibly unique, then we cannot abstract from ourselves what we can (also) share even though we are (also) different. For instance, if I think of myself as irreducibly unique from health to education to economy to fairness, then I cannot think of you as individuals who can share with me something essential from health to education to economy to fairness. Thus, I think that you cannot represent me and my interests and I cannot represent you and your interests: if it is true that we are irreducibly unique, it is also true that our interests are irreducibly contingent, and even idiosyncratic—if it is true that we are irreducibly unique, it is also true that our interests are particulars that cannot lead to universals through abstraction and idealisation, which means that our interests are particulars that cannot lead to laws shared by both of us. What we undermine is precisely the core of representative democracy itself, in that we undermine the idea according to which what rules is something that does not coincide with whatever majority of individuals—conversely, what rules is something that coincides with universals, i.e. laws, as what can be shared by both of us precisely by being abstracted and idealised from both of us. The core of representative democracy itself is the idea according to which ruling is not a prerogative of individuals (otherwise, majorities' contingent, and even idiosyncratic, interests would be absolutised and, consequently, heteronomously imposed on minorities)—the core of representative democracy itself is the idea according to which ruling is a prerogative of laws as universals that, being abstracted and idealised from whatever majority and minority, can be shared by both of them. Which means that, whatever majorities are, they cannot absolutise their contingent, and even idiosyncratic, interests and heteronomously impose them on minorities, in that majorities must also be ruled by laws, starting with common rights and duties, as what must rule over both minorities and majorities. But the idea described above as the core of representative democracy itself is undermined by the kind of individualism we have seen, in that the second logical step of representative democracy's risk is that, if we cannot abstract from ourselves what we can (also) share even though we are (also) different, then we cannot share laws, starting with common rights and duties, as what can rule over both of us. Thus, we fall into the kind of individualism we have seen, which is a solipsism that, first, undermines our intersubjectivity and sense of community both epistemologically and socio-existentially, second, undermines our politics as arising from representative democracy and, third, undermines our commonality by making us risk falling into political anarchism. Actually, if we are ruled by nothing but our contingent, and even idiosyncratic, interests, then we risk falling into political anarchism, in that, if we deprive ourselves of representative democracy as what can make us be represented by experts whenever we lack expertise (for instance, it is not credible that we are experts in health, education, economy and fairness at the same time), then we also deprive ourselves of rulers as laws to which experts contribute in a decisive way—and, if we stop having rulers as laws to which experts contribute in a decisive way, then we also stop having (ἄν) what can actually rule (ἄρχω), which literally means that we fall into political anarchism.

Clearly enough, the more we find examples of political anarchism, the more undermined our intersubjectivity, sense of community, representative democracy and commonality are—which means, as we have seen, that we also increasingly undermine precisely what underpins our capacity to face global, i.e. shared, challenges as the most typical challenges of the unprecedented complexity and uncertainty characterising our era.

In addition to the example of the Italian Five Star Movement, which uses a digital platform to try to move from representative democracy, in which experts work on the drafting of laws, to direct democracy, in which nonexperts also work on the drafting of laws (see Chap. 1), we may think of other examples of political anarchism. More precisely, we may think of other examples in which political anarchism is supported by our use of technology. Since, as we have seen, health is a key issue not only when it comes to addressing humans' core identity but also when it comes to addressing politics, we may start with what follows: "Lies and conspiracy theories about Covid-19, which have amassed millions of views and are accessible to young children, have been available on the social media platform TikTok for months. TikTok accounts with hundreds of thousands of followers that discourage vaccination and peddle myths about Covid survival rates were uncovered by NewsGuard, an organisation that monitors online misinformation. [...] Some of the accounts seen by the Guardian had posted individual videos containing Covid misinformation that had attracted up to 9.2 m views. The misinformation included false comments about side-effects of specific brands of Covid vaccine and misleading comparisons between Covid survival rates and vaccine efficacy rates" (Grierson, Milmo & Farah, 2021[3]). The news reported may be read as an example of a form of totalitarian manipulation. On the one hand, there are children who cannot be as autonomous as adults, in that being capable of self-giving laws (after having distinguished autonomy from heteronomy) requires training, as we have seen, specifically years of training. On the other hand, there are adults who take advantage of necessarily partially heteronomous children, which is a form of totalitarian manipulation that can be nothing but unacceptable from an ethical perspective, in that advantage is taken of the natural imbalance between adults who can be autonomous and children who can be nothing but necessarily partially heteronomous. But extremes frequently happen to converge, and the news reported may be read not only as an example of a form of totalitarianism but also as an example of a form of anarchism for at least the following reason: whenever a society is willing to act towards its own children, who are its future, as reported, a form of anarchism emerges as strongly circulating—whenever a society is willing to act towards its own future as reported, the following anarchic idea emerges as strongly circulating: "anything goes", as Feyerabend writes (as we have seen), in that epistemological and, finally, ethical differences between "Lies", "conspiracy theories", "myths", "misinformation", "false comments" and "misleading comparisons", on the one hand, and at least expert information, on the other hand, dissolve. I consciously write "at least expert information": even though sciences, especially empirical sciences, are not a matter of absolute certainty, as Reichenbach's Hamlet touchingly represents, epistemological and, finally, ethical differences between expert information and nonexpert information are there, starting with differentiating virologists expert in the SARS-CoV-2 virus in particular from doctors in general, as we have seen in Chap. 5.

If we move from children to adults, then the pandemic offers further examples in which political anarchism is supported by our use of technology, starting with the anti-vax movement. It is surely a complex phenomenon to analyse (see at least Berman, 2020) from different perspectives. In what follows, I shall use a strictly philosophical perspective, specifically focused on the relationship between anarchism (as arising from radical individualism), autonomy and technology, specifically social media platforms about wellbeing. The following interviews published in "The Guardian" are interesting cases in point. A first interviewee says: "'I tried to be open-minded and I was open to advice on trying to improve my wellbeing and mental health', she says. So she followed a range of social media accounts, including the 'somatic therapist and biz coach' Phoebe Greenacre, known for her yoga videos, and the 'women's empowerment and spiritual mentor' Kelly Vittengl. The Instagram algorithm did its work. 'I suddenly found myself following so many wellness accounts', she says. When the deployment of the Covid vaccine got under way, [...] [she] began to see posts that troubled her, ranging from polite concern about the social consequences of mass vaccination, or the politics underpinning it, to full-blown rejection of the science. 'The conversation and tone of their posts shifted', she says. 'At first it was

[3] See https://www.theguardian.com/science/2020/jan/27/china-coronavirus-who-to-hold-special-meeting-in-beijing-as-death-toll-jumps (accessed in December 2021).

all about self-care and being part of a community that is caring for each other. But then they started to speak more about how there should be a choice when it came to vaccines. They were saying things like: my body, my choice'. [...] [She] watched as Greenacre posted an Instagram story describing vaccine passport as 'medical apartheid'. Vittengl went further. In a post in July, Vittengl, who is unvaccinated, compared vaccine passports to the social polarisation witnessed during the Holocaust" (Kale, 2021, no page number[4]). A second interviewee increased her success in business, after having become an "'overcoming self-sabotage' coach" (Kale, 2021, no page number) and, especially, an anti-vaxxer: "Before [...] posting anti-vaccine content on her Instagram account, an average post would get 20–30 likes; now, she can easily get more than 150 likes on a post about big pharma. 'The more people get this social reinforcement, the more anti-vaxx they become'" (Kale, 2021, no page number), staff of the Centre for Countering Digital Hate remarks. The interviewer adds that she "sees health through a hyperindividualistic moral frame" (Kale, 2021, no page number). According to the Centre for Countering Digital Hate, "just 12 influencers were responsible for nearly 65% of anti-vaccine content on Facebook and Twitter. 'Many of these leading anti-vaxxers are alternative health entrepreneurs [...]. They're reaching millions of users every day'" (Kale, 2021, no page number), which means increasing their success in business. It is no coincidence that "Technology companies are slow to take down anti-vaccine content, because it is lucrative. [...] Outrage fuels engagement, which drives revenue, for the influencer and the social media platform" (Kale, 2021, no page number). Moreover, the following two points are worth stressing. First, the continuous use of a rhetoric supported by the idea according to which "Outrage fuels engagement": "Comparing vaccine deployment to historic atrocities such as slavery and the Holocaust is a routine trope in anti-vaccine wellness circles" (Kale 2021, no page number). Second, individualism, which even falls into "a hyperindividualistic moral frame", is continuously reported. According to "the co-host of Conspirituality, a podcast about the convergence of conspiracy theories and wellness [...] [,] 'Everything is about personal freedom and personal knowledge. What we see here is late-stage capitalism merging with hyper-individualism'" (Kale, 2021, no page number). Finally, the phenomenon upon which we have reflected through self-tracking technologies and the quantified self emerges: "Doctors are to be distrusted and individuals should take responsibility for their own 'wellness journey'" (Kale, 2021, no page number).

The interviews reported are interesting for several reasons (starting with social media's impact on society in a variety of ways, on which see at least Fuchs, 2014; Lindgren, 2017 and Luttrell, 2021), but my focus is quite precise: the relationship between anarchism (as arising from radical individualism), autonomy and technology. Again, technology is used as a most efficient way to dissolve autonomy (which requires expertise and sense of community, as we have seen) into anarchism as arising from radical individualism (which does not require expertise and sense of

[4] See, also for the following quotes, https://www.theguardian.com/world/2021/nov/11/injecting-poison-will-never-make-you-healthy-how-the-wellness-industry-turned-its-back-on-covid-science (accessed in December 2021).

community, as we have seen). More precisely, social media platforms are used as follows. First, to replace experts with nonexperts: when what matters is "personal knowledge", and not expert knowledge, historians are easily replaced by wellbeing coaches who tell millions of people that there is an analogy between "vaccine passport", on the one hand, and "medical apartheid", "the social polarisation witnessed during the Holocaust" and "historic atrocities such as slavery and the Holocaust", on the other hand. Second, to replace sense of community with radical individualism: when what matters is something sounding "my body, my choice", and not the idea according to which one's body also has to do with one's doctor and one's society, sense of community is easily replaced by "personal freedom" as something absolute that leads to "a hyperindividualistic moral frame", specifically a "late-stage capitalism merging with hyperindividualism". Third, to replace, finally, autonomy with anarchism: when one's instructive comparisons with experts and society dissolve, and "Doctors are to be distrusted and individuals should take responsibility for their own 'wellness journey'", autonomy as acting accordingly to one's own reasons (and not accordingly to others' interests, specifically business interests) and as underpinning one's intersubjectivity is easily replaced by anarchism as the loss of what can actually play a key role in shaping one's autonomy, from distinguishing expert knowledge from nonexpert knowledge to distinguishing intersubjective duties from "hyperindividualistic" rights.

The issue of "hyperindividualistic" rights as arising from anarchism is worth stressing. Several current technologies may be read as supported by the idea, and as equally supporting the idea, according to which we have more and more rights, and even "hyperindividualistic" rights, but less and less duties. Several autonomous technologies seem to primarily show that we have the right to do whatever we want to do, even though we lack expertise, and especially if we lack expertise. For instance, autonomous vehicles seem to primarily show that who cannot drive, from not having driving licence to the young to the elderly to disabled people, have the right to somehow drive—which may be promising from a first ethical perspective, in that it may be a way to promote inclusivity, but worth reflecting from a second ethical perspective, when we realise that the idea according to which rights almost replace duties is exponentially supported by several, and even countless, current technologies. For instance, self-tracking technologies and the quantified self explicitly show that who cannot diagnose, in that they are not doctors, can somehow self-diagnose. The digital platform of the Italian Five Star Movement explicitly shows that who cannot work on the drafting of laws, in that they are not experts, specifically jurists, can somehow work on the drafting of laws. Several softwares explicitly show that who cannot design houses, in that they are not architects, can somehow design houses. Several softwares explicitly show that who cannot write articles, in that they are not writers, can somehow write articles. Several social media platforms explicitly show that who cannot be professionals at doing almost whatever (from cooking to coaching), in that they are amateurs, can somehow be professionals at doing almost whatever (from cooking to coaching). Google explicitly shows that who cannot be professionals at getting whatever kind of information, in that they are not experts, specifically journalists, can somehow get whatever kind of information. And so forth.

Again, the examples described above may be promising (at least sometimes), in that they may be ways to promote inclusivity (at least sometimes). But they are especially worth reflecting, first, for epistemological reasons, in that they are clear symptoms of the kind of crisis of expertise that is readable as epistemological anarchism, as we have seen, and, second, for ethical reasons, in that they are equally clear symptoms of the idea according to which rights almost replace duties—more precisely, the examples described above explicitly show that we exponentially use technology to support the idea according to which rights almost replace duties.

A kind of analogy seems to correlate the epistemological scenario with the ethical scenario. On the one hand, what may be defined as the epistemological articulation of autonomy, i.e. autonomy as acting accordingly to one's own reasons, increasingly collapses, in that one's own reasons increasingly move from being actually one's own reasons, which are underpinned by expert knowledge through instructive comparisons with experts and society, to being actually others' interests, specifically business interests, which are underpinned by nonexpert knowledge. On the other hand, what may be defined as the ethical articulation of autonomy, i.e. autonomy as underpinning one's intersubjectivity, increasingly collapses, in that one's intersubjectivity increasingly moves from being actually one's intersubjectivity, which is underpinned by acknowledging intersubjective rights and duties, to being actually one's individualism, and even "hyperindividualism", which is underpinned by acknowledging one's individualistic, and even "hyperindividualistic", rights, and by not acknowledging one's duties, from the epistemological duty to make the effort to obtain expert knowledge to the ethical duty to make the effort to balance one's rights with others' rights.

Ethical anarchism is the following logical step: the more one acknowledges to have rights, specifically individualistic rights almost replacing duties, the less one acknowledges others' rights as ruling over one's rights and, consequently, duties. It is worth noting that acknowledging others' rights as ruling over one's rights and, consequently, duties does not mean being necessarily heteronomous, but perfectly autonomous. For instance, if one, who is not vaccinated against the SARS-CoV-2 virus, acknowledges others' rights to access hospital therapies for diseases different from Covid-19 (which is difficult because of intensive care occupied by a great number of Covid-19 patients) and acts accordingly, from isolating oneself to being vaccinated, then one is not necessarily heteronomous, but perfectly autonomous accordingly to Kant's categorical imperative. Which means that, if one acknowledges that one's actions, specifically not isolating and not being vaccinated, cannot be universalised (otherwise, healthcare system would collapse), then one is not heteronomously driven by others' reasons, but by reasons autonomously made one's own. Conversely, if one does not acknowledge others' rights (for instance, to access hospital therapies for diseases different from Covid-19) as ruling over one's rights and, consequently, duties, then one falls into ethical anarchism even in the literal sense that what one acknowledges is exclusively the absolute, and not relative, dimension of one's rights, over which what rules (ἄρχω) is nothing (ἄν) but one's senseless idea according to which one, who lives in society, can have individual rights and no social duties at the same time.

And, again, anarchism, specifically ethical anarchism, also means absence of autonomy, specifically ethical autonomy, in that, if the dimension of one's rights is absolutised, then one's actions are equally absolutised in the sense that they do not have actual alternatives from which to choose—and, if one does not have actual alternatives from which to choose, then one does not exercise actual ethical autonomy, in that one's self-given (αὐτός) laws (νόμος) are nothing but what is, randomly, the easiest thing to do accordingly to one's random contingent particularity, and not what is, not randomly at all, the rightest thing to do accordingly to one's nonrandom reasons, which can, and should, move from contingent particularity to universality. For instance, I can move without mindfully choosing, and being driven by, my directions: if I do not face crossroads, and whatever kind of frictional obstacles, then I can continue to move by inertia. Analogously, I can act without mindfully choosing, and being driven by, my reasons: if I do not face alternatives, and whatever kind of frictional obstacles, then I can continue to act by being driven by my random contingent particularity. Which means that, if my intersubjectivity dissolves into ethical anarchism, and, consequently, I lose my capacity to acknowledge others' rights as a kind of frictional obstacles to my rights, then I stop exercising my ethical autonomy as my capacity to choose from different reasons, in that I have nothing but my random contingent particularity that, being absolutised, i.e. deprived of alternatives, cannot move from randomness to something that can actually be defined as self-given laws.

The circle of our reflection upon our relationship with technology, specifically the relationship between our autonomy and technology, seems to close. Again, we seem to increasingly become analogous to Kant's "*automaton*" "*spirituale*", which means that, again, we seem to increasingly become analogous to "turnspit[s]" (see Chap. 2)—more precisely, our autonomy, together with our freedom as underpinned by it, seems to increasingly become analogous to "the freedom of a turnspit, which, when once it is wound up, also accomplishes its movements of itself", in that, "if the freedom of our will were none other than [...] psychological and comparative but not also transcendental [...], then it would at bottom be nothing better than the freedom of a turnspit".

And what matters most, at least from the perspective of our reflection, is that, if what we actually want, sometimes consciously and sometimes unconsciously, is to unburden ourselves of our autonomy as what has increasingly become our most unbearable burden in the last three centuries, then technology may be our best solution—technology may be our best solution to make our autonomy dissolve into anarchism as its antonym: current technologies show our move from autonomy to anarchism both as immanentisations of rulers in general and as immanentisations of the divine in particular (which may be our most radically anarchic move, as we have seen).

Could we imagine moves more radically anarchic than the following series of immanentisations of the divine? It is no coincidence that the following series seems to be a series of miracles. Even though we cannot drive, we can somehow drive. Even though we cannot diagnose, we can somehow self-diagnose. Even though we cannot work on the drafting of laws, we can somehow work on the drafting of laws. Even though we cannot design houses, we can somehow design houses. Even though we cannot write articles, we can somehow write articles. Even though we cannot be professionals, we can somehow be professionals. And so forth. And the reason why we can somehow drive, self-diagnose, work on the drafting of laws, design houses, write articles, be professionals and so forth is that we create technologies that make us do almost whatever without exercising our own actual ἀρχή (transliterated as *arche*), our own actual νόμος—more precisely, we create technologies that replace our own actual ἀρχή, our own actual νόμος by being (immanently) omnipresent (they are everywhere, i.e. always with us, from our desks to our pockets), omniscient (they know everything, i.e. not only the answers to our questions but also ourselves by tracking us), omnipotent (they have more and more power over us, from shaping our worldviews to shaping our decisions and, consequently, our actions) and inscrutable (by trading their automation for a kind of autonomy, they make us acknowledge their outputs through a kind of act of faith).

We may seem victims of a process of which we increasingly lose control, but we are far from being victims—our being active, sometimes consciously and sometimes unconsciously, and not passive, is, at least, one of my arguments, according to which we purposely create technologies as perfect scapegoats, i.e. our own autonomy's perfect avatars, through which, from when we should find "the courage I need for my action" to when we should face its (irreducibly uncertain) outcomes, we unburden ourselves of our heaviest burdens, from our sense of responsibility, in the best case scenario, to our sense of guilt, in the worst case scenario.

And, if my argument makes sense, then the possible answers to my two questions may be the following. As far as the first question is concerned (if our era, characterised by unprecedented complexity and uncertainty, may be the worst time to lose not only our autonomy in general but also our autonomy as underpinning our intersubjectivity in particular, precisely as what finally underpins our capacity to face global, i.e. shared, challenges, then what may be the philosophical meaning of the paradoxical overturning of our autonomy?), its possible answer may be that, after autonomy's irresistible rise especially from modern culture to contemporary culture, our autonomy has become our most unbearable burden not only when it comes to facing our responsibilities for our own lives (when our decisions and actions impact our own lives) but also when it comes to facing our responsibilities for others' lives, i.e. our social responsibilities (when our decisions and actions impact others' lives)—which is precisely what our global, i.e. shared, challenges increasingly demand of us. Consequently, as far as the second question is concerned (what may be, from an ethical perspective, the philosophical meaning of our paradoxical trade of our autonomy for technology's automation?), its possible answer may be that, even more precisely, our perfect scapegoat, i.e. our own autonomy's perfect avatar, is automation itself, as the kind of randomness that unburdens us of whatever kind of responsibility, from responsibilities for our own lives to responsibilities for others' lives, by replacing the relationship between (autonomous) reasons and (autonomous) actions with the relationship between (automated) causes and (automated) effects—automation itself is our perfect scapegoat, i.e. our own autonomy's perfect avatar, by unburdening us of our responsibilities both as single individuals and as a society.

[illegible] the [illegible] [illegible] the following [illegible] [illegible] [illegible] and [illegible] may be [illegible] [illegible] [illegible] [illegible] [illegible] [illegible] [illegible] [illegible] [illegible] [illegible] [illegible] [illegible] [illegible] [illegible] only when [illegible] [illegible] [illegible] and actions [illegible] [illegible] [illegible] responsibilities [illegible] [illegible] [illegible] and [illegible] [illegible] [illegible] which [illegible] [illegible] [illegible] [illegible] [illegible] [illegible] [illegible] [illegible] [illegible] [illegible] [illegible] [illegible] [illegible] meaning of [illegible]

Conclusion

The core of my work is arguing that we are active in dramatically, and even touchingly, trying to get by, by purposely creating technologies as perfect scapegoats, i.e. our own autonomy's perfect avatars, through which, from when we should find the courage to act to when we should face our actions' irreducibly uncertain outcomes, we unburden ourselves of our heaviest burdens, from our sense of responsibility, in the best case scenario, to our sense of guilt, in the worst case scenario.

If the core of my work makes sense, then our technological era actually challenges one of the most essential cornerstones of the definition itself of human identity and dignity in Western culture, in that our technological era is characterised by an increasing loss of rules, which are increasingly replaced by a kind of automated randomness that may be read as the most radical form of anarchism we have ever experienced in Western culture—and losing rules, which dissolve into a kind of immanentised automated randomness, specifically technology's, means losing autonomy, from both an etymological perspective and a philosophical perspective.

The first words of my work are about the capacity from which philosophy itself arises—the capacity to "connect" the dots through a kind of synthesis, i.e. vision of the whole, after having analysed them. Thus, the conclusion of my work may also be an initial attempt to "connect" its outcomes with a further greater scenario of relationships. If my work is a matter of trying to "connect" current technologies, first, with dots as historical cornerstones that can make more sense of them and, second, and especially, with dots as philosophy's classic questions that can go even further in making more sense of them, starting with what the relationship between current technologies and human autonomy is, then the conclusion of my work may also be a matter of trying to initially "connect" the core of my answer with a further greater scenario of relationships—more precisely, the conclusion of my work may also be an initial attempt to "connect" the idea according to which the relationship between current technologies and human autonomy is a matter of overturning, i.e. trading human autonomy for technological automation, with dots as a further greater scenario of relationships.

S. Chiodo, *Technology and the Overturning of Human Autonomy*, Studies in Applied Philosophy, Epistemology and Rational Ethics 66,
https://doi.org/10.1007/978-3-031-26159-6

The clue may be the final idea of the first part of my work, according to which power of creation seems to play a key role in our technological era. If it is true that technology seems to increasingly occupy whatever ontological dimension, from human ontology to divine ontology, in addition to technological ontology, it is also true that the phenomenon described above may be read, on the one hand, as a kind of creators' surrender (since their burdens have become heaviest, and even unbearable, they create technology as their best scapegoat to unburden themselves of them) and, on the other hand, as a kind of creators' success (they may prove their power precisely by creating an even more powerful creation, i.e. technology, to which to finally surrender)—paradoxically enough, as the history of the figure of Prometheus also hints, surrendering to one's own creation may be one's own power's, specifically power of creation's, best proof. More precisely, as we have seen, our technological "exosel[ves]" may be read, on the one hand, as powerful to the point that they somehow create our selves and, on the other hand, as still our creations—our technological "exosel[ves]" are still our perfectly immanent artefactual creations. Thus, again, creation, specifically power of creation, plays a key role—power of creation exceedingly marks our technological era, if it is true that the latter may be read as a matter of exploiting the former's most pervasive outcome, i.e. technology, as our most powerful way to try to (dis)solve our most dramatic burdens by making technology be our best scapegoat.

If we ask ourselves where the notion of creation, specifically humans' power of creation, starts its history, at least in Western culture, then we should answer that, even though its genesis is ancient (for instance, according to Plato, "For of anything whatever that passes from not being into being the whole cause is composing [ποίησις, transliterated as *poiesis*]", Plat. *Sym.* 205 b-c), its rise is modern and contemporary, specifically Romantic (see at least Tatarkiewicz, 1975), when the notion of creation as it is meant by the Latin word "*creatio*" (which has to do with creating something from nothing) surpasses the notion of creation as it is meant by the Latin word "*inventio*" (which has to do with creating something from something. See also Steiner, 1990). Again, an overturning of hierarchical differences emerges: humans start moving from being represented as mastering creation as *inventio* to being represented as mastering creation as *creatio*, which typically characterises the divine as one of its prerogatives. But a further overturning of hierarchical differences seems to emerge: technology, specifically current technologies upon which we have reflected, seems to start moving from being thought of as a matter of creation as *inventio* to being thought of as a matter of creation as *creatio* for at least two reasons, if it is true that technology seems to increasingly occupy whatever ontological dimension, from human ontology to divine ontology, in addition to technological ontology. First, technology as something created starts getting out from under humans as its own creators (it is no coincidence that the issue of technology as getting out from under humans as its own creators especially arises from Romantic culture, as we have seen in the case of the modern Prometheus). Moreover, it is no coincidence that getting out from under humans as technology's creators means being somehow autonomous—technology also seems to start being somehow autonomous by moving from being a creation implying the stronger control of *inventio* to being a creation implying the less strong control of *creatio*. In the first case, i.e. *inventio*, creating from something that is necessarily known may imply that creators have more control over creations, in that the former have elements to predict the latter's destiny more easily. In the second case, i.e. *creatio*, creating from nothing may imply that creators have less control over creations, in that the former do not have elements to predict the latter's destiny more easily. Second, technology as something that can create starts being even more unpredictable, which is something that, again, makes technology somehow move from creation as *inventio* to creation as *creatio*. For instance, when we have reflected upon current technologies both as increasingly omnipotent and as increasingly inscrutable, we have defined them as having more and more power over us, from shaping our worldviews to shaping our decisions and, consequently, our actions (and we may somehow think of shaping us as a kind of creation as *creatio*, in that, even though we exist before current technologies shape us, our digital selves may be thought of as ontologically different from our analog selves, as we have seen). In addition, we have reflected upon algorithms that, as the basis of a great number of other current technologies, can be defined as black boxes, in that the reasons why a given input results in a given output can be inscrutable (and we may somehow think of outputs' inscrutability as correlating with the kind of unpredictability characterising creation as *creatio* more than creation as *inventio*, in that the process of creating moves from being a matter of effects of causes that are known to being a matter of effects of causes that are not known and, consequently, mean nothing to us and somehow dissolve into a kind of nothing for us).

Thus, surprisingly enough, several Romantic notions seem to characterise current technologies: not only the challenge between humans and the divine (which arises from ancient culture and exceedingly develops in modern and contemporary culture[1]) and radical individualism from which anarchism arises,[2] as we have seen, but also creation, specifically power of creation, as *creatio*. Should we think that our technological era may be promisingly read as a last, and even as the last, outcome of Romantic culture? Differences between our technological era and Romantic culture also emerge, starting with the former's focus on *logos* against the latter's focus on other forms of rationality, and even on forms of irrationality, in addition to *logos*. Yet, there is a meaningful sense in which our technological era's focus on *logos* and Romantic culture's focus on other ways to obtain reliable knowledge may correlate: in both cases, knowledge may be what humans want to obtain as their way to finally overturn the most crucial hierarchies defining them, from power (between rulers and ruled) to ontology (between creators and created), as we have seen.

---

[1] As far as Romantic culture in particular is concerned, see at least the notion of Titanism especially arising from Goethe (1774, 1790 and 1819), which is meaningfully his *Prometheus* ("Cover your heaven, Zeus, / with cloudy vapours, / and test your strength, like a boy / beheading thistles, / on oaks and mountain peaks; / yet you must leave / my earth alone, / and my hut you did not build, / and my hearth, / whose fire / you envy me. [...] Here I sit, making men / in my own image, / a race that shall be like me, / that shall suffer, weep, / know joy and delight, / and ignore you / as I do!", Goethe, 1819: verses 1-11 and 51-57, translated ds.). R. Stokes).

[2] As far as Romantic culture in particular is concerned, see at least the notion of egotism especially arising from Stendhal (1832).

More precisely, in both case, knowledge, especially applied knowledge as humans' creations, from current technologies to Romantic arts, may be what humans want to obtain as their way to finally overturn their being ruled and created: first, they aspire to the privilege of becoming rulers and creators by applying knowledge to create especially current technologies and, second, they aspire to the (paradoxical) privilege of somehow becoming, again, ruled and created especially by current technologies as their own autonomy's perfect avatars—humans aspire to the (paradoxical) privilege of somehow becoming, again, ruled and created especially by current technologies through which, from when they should find the courage to act to when they should face their actions' irreducibly uncertain outcomes, they unburden themselves of their heaviest burdens, by making current technologies be their perfect scapegoats as the perfect avatars of their own autonomy's heaviest burdens, from sense of responsibility to sense of guilt.

If my working hypothesis makes sense, then what may be our future relationship with technology? History teaches us that extreme phenomena are frequently followed by equally extreme phenomena. Predictions of the future cannot be the aim of my work. Yet, if one believes that it is worth trying, at least, to add wisdom to *logos*, specifically the former's horizontality and farsightedness to the latter's verticality and shortsightedness, in that we desperately need the capacity to imagine especially what we cannot know as at least probabilistically "true", "certain", because of our equally desperate need to have reasons to act that can be valid even though they cannot be "true", "certain", then starting reflecting may not be senseless. Thus, where to start? A possible starting point may be the following: if it is true that extreme phenomena are frequently followed by equally extreme phenomena, it is also true that we should bear in mind that power of creation's antonym is sterility, of which we may variously think. Sterility may be technology's, if it turns out to continue to lack imagination underpinning one's (something's?) own reasons to make sense of one's (something's?) own decisions, actions and, finally, lives. Sterility may be humans', if they turn out to continue to atrophy their most creative capacities, from their capacity to imagine to their capacity to shape, especially when they need to face uncertainty. Sterility may be the divine's, if it turns out to continue to be thought of as dissolving into a kind of immanentised automated randomness, specifically technology's (and together with humans' autonomy). And sterility may be nature's, if it turns out to continue to be thought of as a kind of technology's, humans' and the divine's counterpart to challenge, and not as something to which all of them are somehow tied.

References

Aeschylus. (1926). *Prometheus bound (PB)* (H. W. Smith, Trans.). Harvard University Press.
Ajana, B. (Ed.). (2018). *Metric culture. Ontologies of self-tracking practices*. Emerald Publishing Limited.
Ajana, B. (Ed.). (2018). *Self-tracking. Empirical and philosophical investigations*. Springer.
Ajana, B. (2020). Personal metrics. Users' experiences and perceptions of self-tracking practices and data. *Social Science Information, 4*(59), 654–678.
Alpaydin, E. (2016). *Machine learning. The new AI*. MIT Press.
Amoroso, D., & Tamburrini, G. (2021). In search of the "human element". International debates on regulating autonomous weapons systems. *The International Spectator, 1*(56), 20–38.
Anderson, M., & Anderson, S. L. (2011). *Machine ethics*. Cambridge University Press.
Aristotle. (1934). *Nicomachean ethics* (H. Rackham, Trans.). Harvard University Press.
Aristotle. (1991). *The art of rhetoric* (J. H. Freese, Trans.). Harvard University Press.
Arthur, W. B. (2009). *The nature of technology. What it is and how it evolves*. Free Press.
Asimov, I. (1956). *I, robot*. The New American Library.
Augustine. (2012). Confessions. In J. J. O'Donnell (Ed.). Oxford University Press.
Awad, E., et al. (2018). The moral machine experiment. *Nature, 563*, 59–75.
Awad, E., et al. (2020a). Crowdsourcing moral machines. *Communications of the ACM, 3*(63), 48–55.
Awad, E., et al. (2020b). Universals and variations in moral decisions made in 42 countries by 70.000 participants. *Proceedings of the National Academy of Sciences, 5*(117), 2332–2337.
Badura, C., & Kind, A. (2021). *Epistemic uses of imagination*. Routledge.
Banerjee, S. (2021). Autonomous vehicles. A review of the ethical, social and economic implications of the AI revolution. *International Journal of Intelligent Unmanned Systems, 4*(9), 302–312.
Barabás, I., Todoruţ, A., Cordoş, N., & Molea, A. (2017). Current challenges in autonomous driving. *IOP Conference Series. Materials Science and Engineering, 252*, 1–6.
Baricco, A. (2018). *The game*. Einaudi.
Baron, C., Halvorsen, P. N., & Cornea, C. (Eds.). (2017). *Science fiction, ethics and the human condition*. Springer.

S. Chiodo, *Technology and the Overturning of Human Autonomy*, Studies in Applied Philosophy, Epistemology and Rational Ethics 66,
https://doi.org/10.1007/978-3-031-26159-6

Barredo Arrieta, A., et al. (2020). Explainable artificial intelligence (XAI). Concepts, taxonomies, opportunities and challenges toward responsible AI. *Information Fusion, 58*, 82–115.

Bellet, T., et al. (2019). From semi to fully autonomous vehicles. New emerging risks and ethico-legal challenges for human-machine interactions. *Transportation Research Part F. Traffic Psychology and Behaviour, 63*, 153–164.

Belli, J. (2016). Unhappy? There's an app for that. Tracking well-being through the quantified self. *Digital Culture and Society, 1*(2), 89–103.

Benson, P. (1994). Autonomy and self-worth. *The Journal of Philosophy, 12*(91), 650–668.

Berg Olsen, J. K., & Voetmann Christiansen, F. (2009). Technology and science epistemology, rationality and the empirical turn. *Synthese, 3*(168), 313–318.

Berlin, I. (1958). *Two concepts of freedom*. Oxford University Press.

Berlin, I. (1969). *Four essays on liberty*. Oxford University Press.

Berlin, I. (1999). *The roots of romanticism*. Princeton University Press.

Berlin, I. (2000). *The power of ideas*. Chatto & Windus.

Berlin, I. (2006). *Political ideas in the romantic age. Their rise and influence on the modern thought*. Princeton University Press.

Berlin, I. (1997). Two concepts of liberty (1958). In *The proper study of mankind* (pp. 191–242). Chatto & Windus.

Berlin, I. (1988). *The pursuit of the ideal*. The New York Review of Books.

Berlin, I. (1990). *The crooked timber of humanity. Chapters in the history of ideas*. Murray.

Berlin, I. (1996). *The sense of reality. Studies in ideas and their history*. Chatto & Windus.

Berman, J. M. (2020). *Anti-vaxxers. How to challenge a misinformed movement*. MIT Press.

Berofsky, B. (1995). *Liberation from self. A theory of personal autonomy*. Cambridge University Press.

Biondo, N., & Canestrari, M. (2019). *Il sistema Casaleggio. Partito, soldi, relazioni: Ecco il piano per manomettere la democrazia*. Ponte alle Grazie.

Blomseth Christiansen, T. *Debugging my allergy*. https://quantifiedself.com/show-and-tell/?project=563

Blumenberg, H. (1985). *Work on myth* (R. M. Wallace, Trans.). MIT Press.

Böhme, G. (2012). *Invasive technification. Critical essays in the philosophy of technology* (C. Shingleton, Trans.). Bloomsbury.

Bonnefon, J. F., et al. (2016). The social dilemma of autonomous vehicles. *Science, 352*, 1573–1576.

Bostrom, N. (2014). *Superintelligence. Paths, dangers, strategies*. Oxford University Press.

Bouhours, D. (1671). *Les entretiens d'Ariste et d'Eugène*. Chez Sebastian Marbre-Cramoisy.

Bouk, D. (2015). *How our days became numbered*. University of Chicago Press.

Bowker, G. C., & Leigh Star, S. (1999). *Sorting things out. Classifications and its consequences*. MIT Press.

Bozdag, E., & van den Hoven, J. (2015). Breaking the filter bubble. Democracy and design. *Ethics and Information Technology, 17*, 249–265.

Bradshaw-Martin, H., & Easton, C. (2014). Autonomous or "driverless" cars and disability. A legal and ethical analysis. *European Journal of Current Legal Issues, 3*(20), 1–17.

Bratman, M. E. (2007). *Structures of agency. Essays*. Oxford University Press.

Brende, E. (2005). *Better off. Flipping the switch on technology*. Harper Perennial.

Brogård Kristensen, D., & Ruckenstein, M. (2018). Co-evolving with self-tracking technologies. *New Media & Society, 10*(20), 3642–3640.

Brontë, C. (1971). Jane Eyre. In R. J. Dunn (Ed.), *An authoritative text, backgrounds, criticism* (1847). Norton.

Brynjolfsson, E., & McAfee, A. (2014). *The second machine age. Work, progress and prosperity in a time of brilliant technologies*. Norton & Co.

Brynjolfsson, E., & McAfee, A. (2017). *Machine, platform, crowd. Harnessing our digital future*. Norton & Co.

Brzechczyn, K. (Ed.). (2009). Idealization XIII. Modeling in history. *Poznań Studies in the Philosophy of the Sciences and the Humanities, 97*.

Brzeziński, J., & Nowak, L. (Eds.). (1992). Idealization III. Approximation and truth. *Poznań Studies in the Philosophy of the Sciences and the Humanities, 25.*

Brzeziński, J., Coniglione, F., Kuipers, T. A. F., & Nowak, L. (Eds.). (1990a). Idealization I. general problems. *Poznań Studies in the Philosophy of the Sciences and the Humanities, 16.*

Brzeziński, J., Coniglione, F., Kuipers, T. A. F., & Nowak, L. (Eds.). (1990b). Idealization II. forms and applications. *Poznań Studies in the Philosophy of the Sciences and the Humanities, 17.*

Calvert, S. C., Heikoop, D. D., Mecacci, G., & van Arem, B. (2020). A human centric framework for the analysis of automated driving systems based on meaningful human control. *Theoretical Issues in Ergonomics Science, 4*(21), 478–506.

Cambra-Badii, I., Guardiola, E., & Banos, J.-E. (2020). The ethical interest of Frankenstein. Or, the modern Prometheus. A literature review 200 years after its publication. *Science and Engineering Ethics, 5*(26), 2791–2808.

Carmichael, A. (2010). *Why I stopped tracking.* https://quantifiedself.com/blog/why-i-stopped-tracking/

Carr, N. G. (2014). *The glass cage. Automation and us.* Norton & Co.

Carruthers, P. (2009). How we know our own minds. The relationship between mindreading and metacognition. *Behavioral and Brain Sciences, 2*(32), 121–138.

Cartwright, N., & Jones, M. R. (Eds.). (2005). Idealization XII. Correcting the model. Idealization and abstraction in the sciences. *Poznań Studies in the Philosophy of the Sciences and the Humanities, 86.*

Cassirer, E. (1923). Einstein's theory of relativity. In (W. C. Swabey & M. C. Swabey, Trans.). *Substance and function and Einstein's theory of relativity.* Open Court.

Cheney-Lippold, J. (2011). A new algorithmic identity. Soft biopolitics and the modulation of control. *Theory, Culture and Society, 6*(28), 164–181.

Chiodo, S., & De Cristofaro, D. (2023). Quantified sleep. Self-tracking technologies and the reshaping of 21st-century subjectivity. *Historical Social Research.*

Chiodo, S. (2023). Trading human autonomy for technological automation. In S. Lindgren (Ed.), *Handbook of critical studies of artificial intelligence.* Edward Elgar Publishers.

Chiodo, S. (2020a). *Technology and anarchy. A reading of our era.* Lanham-Boulder-New Lexington Books-The Rowman & Littlefield Publishing Group.

Chiodo, S. (2020b). The greatest epistemological externalisation. Reflecting on the puzzling direction we are heading to through algorithmic automatisation. *AI & Society. Journal of Knowledge, Culture and Communication, 2*(35), 431–440.

Chiodo, S. (2021). From je ne sais quoi to quantified self. A philosophical agenda. *Studi di Estetica, 3,* 157–175.

Chiodo, S. (2022a). Adding wisdom to computation. The task of philosophy today. *Metaphilosophy, 1*(53), 70–84.

Chiodo, S. (2022b). Human autonomy, technological automation (and reverse). *AI & Society. Journal of Knowledge, Culture and Communication, 1*(37), 39–48.

Chiodo, S. (2022c). The quantified self as epistemological anarchism. *Philosophia. Philosophical Quarterly of Israel, 4*(50), 1665–1685.

Christman, J. (Ed.). (1989). *The inner citadel. Essays on individual autonomy.* Oxford University Press.

Christman, J. (2001). *The politics of persons. Individual autonomy and socio-historical selves.* Cambridge University Press.

Christman, J., & Anderson, J. (Eds.). (2005). *Autonomy and the challenges to liberalism.* Cambridge University Press.

Christman, J. (1991). Autonomy and personal history. *Canadian Journal of Philosophy, 21,* 1–24.

Clark, A., & Chalmers, D. (1998). The extended mind. *Analysis, 1*(58), 7–19.

Clark, A. (2003). *Natural-born cyborgs. Minds, technologies and the future of human intelligence.* Oxford University Press.

Coeckelbergh, M. (2009). Personal robots, appearance, and human good. A methodological reflection on roboethics. *International Journal of Social Robotics, 1*(3), 217–221.

Coeckelbergh, M. (2016). Responsibility and the moral phenomenology of using self-driving cars. *Applied Artificial Intelligence. An International Journal*, *8*(30), 748–757.
Coeckelbergh, M. (2017). *New romantic cyborgs. Romanticism, information technology, and the end of the machine*. MIT Press.
Coeckelbergh, M. (2020a). *AI ethics*. MIT Press.
Coeckelbergh, M. (2020b). Artificial intelligence, responsibility attribution, and a relational justification of explainability. *Science and Engineering Ethics*, *26*, 2051–2068.
Collins, H. M. (2014). *Are we all scientific experts now?* Polity Press.
Collins, H. M., & Evans, R. (2007). *Rethinking expertise*. University of Chicago Press.
Collins, H. M., & Evans, R. (2002). The third wave of science studies. Studies of expertise and experience. *Social Studies of Science*, *2*(32), 235–296.
Confalonieri, R., Coba, L., Wagner, B., & Besold, T. R. (2021). A historical perspective of explainable artificial intelligence. *WIREs Data Mining and Knowledge Discovery*, *11*, 1–21.
Coniglione, F., Poli, R., & Rollinger, R. (Eds.). (2004). Idealization XI. Historical studies on abstraction and idealization. *Poznań Studies in the Philosophy of the Sciences and the Humanities*, *82*.
Contissa, G., Lagioia, F., & Sartor, G. (2017). The ethical knob. Ethically-customisable automated vehicles and the law. *Artificial Intelligence and Law*, *25*, 365–378.
Corbetta, P., & Gualmini, E. (Eds.). (2013). *Il partito di Grillo*. Il mulino.
Crawford, K., Lingel, J., & Karppi, T. (2015). Our metrics, ourselves. A hundred years of self-tracking from the weight scale to the wrist wearable. *European Journal of Cultural Studies*, *18*, 479–496.
Crawford, K. (2021). *Atlas of AI. Power, politics, and the planetary costs of artificial intelligence*. Yale University Press.
Cressey, D. (2017). The DIY electronics transforming research. *Nature*, *544*(7648), 125–126.
Cunneen, M., Mullins, M., & Murphy, F. (2019). Autonomous vehicles and embedded artificial intelligence. The challenges of framing machine driving decisions. *Applied Artificial Intelligence*, *8*(33), 706–731.
Cunneen, M., et al., (2020). Autonomous vehicles and avoiding the trolley (dilemma). Vehicle perception, classification, and the challenges of framing decision ethics. *Cybernetics and Systems*, *1*(51), 59–80.
Currie, A. (Ed.). (2018). Futures of research in catastrophic and existential risk. *Futures*, *102*.
Cuypers, S. (2001). *Self-identity and personal autonomy*. Ashgate.
D'Angelo, P., & Velotti, S. (Eds.). (1997). *Il "non so che". Storia di una idea estetica*. Aesthetica Edizioni.
Dalal, N., Intezari, A., & Heitz, M. (Eds.). (2016). *Practical wisdom in the age of technology. Insights, issues, and questions for a new millennium*. Routledge.
Darwall, S. (2006). The value of autonomy and autonomy of the will. *Ethics. An International Journal of Social, Political, and Legal Philosophy*, *2*(116), 263–284.
Dastin, J. (2018). *Amazon scraps secret AI recruiting tool that showed bias against women*. https://www.reuters.com/article/us-amazon-com-jobs-automation-insight/amazon-scraps-secret-ai-recruiting-tool-that-showed-bias-against-women-idUSKCN1MK08G
De Groot, M., Drangsholt, M., Martin-Sanchez, F. J., & Wolf, G. I. (2017). Single subject (N-of-1) research design, data processing, and personal science. *Methods of Information in Medicine*, *6*(56), 416–418.
Deligiorgi, K. (2012). *The scope of autonomy*. Oxford University Press.
Descartes, R. (1912). *A discourse on method* (1637) (J. Veitch, Trans.). J.M. Dent and Sons.
Descartes, R. (1983). *Principles of philosophy* (1644). In V. R. Miller & R. P. Miller (Eds.). Reidel.
Descartes, R. (1985). *Preface* to the French edition of *Principles of philosophy* (1647). In C. Adam & P. Tannery (Eds.), *Oeuvres de Descartes*, Paris, Vrin, 1964–76. English translation in J. Cottingham, R. Stoothoff, D. Murdoch (Eds.), *The philosophical writings of Descartes*. Cambridge University Press.

Descartes, R. (1958). Passions of the soul (1649). In C. Adam & P. Tannery (Eds.), *Oeuvres de Descartes*. Paris, Vrin, 1964–76. English translation in J. Cottingham, R. Stoothoff, D. Murdoch (Eds.), *The philosophical writings of Descartes*. Cambridge University Press.

Descartes, R. (1985). Rules for the direction of the mind (1684). In C. Adam & P. Tannery (Eds.), *Oeuvres de Descartes*, Paris, Vrin, 1964–76. English translation in J. Cottingham, R. Stoothoff, D. Murdoch (Eds.), *The philosophical writings of Descartes*. Cambridge University Press.

Dickenson, D. (2013). *Me medicine vs. we medicine. Reclaiming biotechnology for the common good*. Columbia University Press.

Dietrich, M., & Weisswange, T. H. (2019). Distributive justice as an ethical principle for autonomous vehicle behavior beyond hazard scenarios. *Ethics of Information Technology, 21*, 227–239.

Domingos, P. (2015). *The master algorithm. How the quest for the ultimate learning machine will remake our world*. Allen Lane.

Dougherty, C. (2006). *Prometheus*. Taylor & Francis.

Dryden, J. *Autonomy*. Available at https://iep.utm.edu/autonomy/

Duhigg, C. (2013). *The power of habit. Why we do what we do and how to change*. Random House.

Durbin, P., & Rapp, F. (Eds.). (1980). *Philosophy and technology*. Reidel.

Dusenbery, M. (2018). *Doing harm. The truth about how bad medicine and lazy science leave women dismissed, misdiagnosed, and sick*. Harper Collins.

Dworkin, G. (1988). *The theory and practice of autonomy*. Cambridge University Press.

Dyson, G. B. (1997). *Darwin among the machines. The evolution of global intelligence*. Penguin.

Edel, A. (1945). The evaluation of ideals. *The Journal of Philosophy, 21*(42), 561–577.

Eggers, D. (2014). *The circle. A novel*. Vintage.

Ekstrom, L. W. (1993). A coherence theory of autonomy. *Philosophy and Phenomenological Research, 53*, 599–616.

Emmet, D. (1994). *The role of the unrealisable. A study in regulative ideals*. St. Martin's Press.

Epictetus (1800). The discourses of Epictetus. In G. Long (Ed.). Estes & Lauriat.

Epting, S. (2019). Automated vehicles and transportation justice. *Philosophy & Technology, 32*, 389–403.

Ess, C. (2016). Phronesis for machine ethics? Can robots perform ethical judgments? *Frontiers in Artificial Intelligence and Applications, 290*, 386–389.

European Technology Platform on Smart Systems Integration (2015). *European roadmap. Smart systems for automated driving*.

Faulhaber, A. K. et al. (2019). Human decisions in moral dilemmas are largely described by utilitarianism. Virtual car driving study provides guidelines for autonomous driving vehicles. *Science and Engineering Ethics, 25*, 399–418.

Federal Minister of Transport and Digital Infrastructure (2017). *Ethics commission. Automated and connected driving*. Report (extract).

Ferretti, F. (2019). Mapping do-it-yourself science. *Life Sciences, Society and Policy, 15*(1), 1–23.

Feyerabend, P. K. (1987). *Farewell to reason*. Verso.

Feyerabend, P. K. (1975). *Against method. Outline of an anarchistic theory of knowledge*. NLB.

Fiala, A. (2013). *Against religion, wars, and states. the case for enlightenment atheism, just war pacifism, and liberal-democratic anarchism*. The Rowman & Littlefield Publishing Group.

Fiala, A. (2017). Anarchism. In *Stanford encyclopaedia of philosophy*. https://plato.stanford.edu/entries/anarchism/#TheoAnar

Fichte, J. G. (1993). Some lectures concerning the scholar's vocation (1794). In D. Breazeale (Ed.), *Early philosophical writings*. Cornell University Press.

Fichte, J. G. (1982). Foundations of the entire science of knowledge (1794–95). In P. Heath & J. Lachs (Eds.), *Fichte. Science of knowledge (Wissenschaftslehre)*. Cambridge University Press.

Finn, E., Guston, D., & Robert, J. S. (Eds.). (2017). *Frankenstein. Or, the modern Prometheus. Annotated for scientists, engineers, and creators of all kinds*. MIT Press.

Finn, E. (2017). *What algorithms want. Imagination in the age of computing*. MIT Press.

Fischer, J. M., & Ravizza, M. (1998). *Responsibility and control*. Cambridge University Press.

Floridi, L. (2013). *The ethics of information*. Oxford University Press.

Floridi, L. (2014). *The 4th revolution*. Oxford University Press.
Floridi, L. (2008). Artificial intelligence's new frontier. Artificial companions and the fourth revolution. *Metaphilosophy, 4–5*(39), 651–655.
Floridi, L. (2019). Autonomous vehicles. From whether and when to where and how. *Philosophy & Technology, 4*(32), 569–573.
Fossa, F. (2022). *Ethics of driving automation. Artificial agency and human values*. Springer.
Foucault, M. (1997). Technologies of the self. In P. Rabinow (Ed.), *Ethics. Subjectivity and truth* (pp. 223–252). The New Press.
Frankel Paul, E., Miller, F., & Paul, J. (Eds.). (2003). *Autonomy*. Cambridge University Press.
Frankfurt, H. (Ed.). (1988). *The importance of what we care about*. Cambridge University Press.
Frankfurt, H. (1999). *Necessity, volition, and love*. Cambridge University Press.
Frankfurt, H. (1971). Freedom of the will and the concept of a person. *The Journal of Philosophy, 1*(68), 5–20.
Franks, B., & Wilson, M. (Eds.). (2010). *Anarchism and moral philosophy*. Palgrave Macmillan.
Franssen, M., Kroes, P., Reydon, T. A. C., & Vermaas, P. E. (2014). *Artefact kinds. Ontology and the human-made world*. Springer.
Franssen, M., Lokhorst, G.-J., & van de Poel, I. (2018). Philosophy of technology. In *Stanford encyclopaedia of philosophy*. https://plato.stanford.edu/archives/fall2018/entries/technology/
Fridman, L. *Human-centered autonomous vehicle systems. Principles of effective shared autonomy* (pp. 1–9). arXiv:1810.01835
Friedman, M. (2003). *Autonomy, gender, politics*. Oxford University Press.
Frischmann, B., & Selinger, E. (2018). *Re-engineering humanity*. Cambridge University Press.
Fuchs, C. (2014). *Social media. A critical introduction*. Sage.
Galilei, G. (1967). *Dialogue concerning the two chief world systems Ptolemaic and Copernican*. In S. Drake (Ed.). University of California Press.
Gans, C. (1992). *Philosophical anarchism and political disobedience*. Cambridge University Press.
Garreau, J. (2006). *Radical evolution. The promise and peril of enhancing our minds, our bodies, and what it means to be human*. Broadway Books.
Gaut, B. (2007). Creativity and imagination. In B. Gaut & P. Livingston (Eds.), *The creation of art. New essays in philosophical aesthetics* (pp. 148–173). Cambridge University Press.
Gerdes, J. C., & Thornton, S. M. (2015). Implementable ethics for autonomous vehicles. In E. Maurer et al., (Eds.), *Autonomous driving. Technical, legal and social aspects* (pp. 87–102). Springer.
Gerdes, A. (2016). The role of phronesis in robot ethics. *Frontiers in Artificial Intelligence and Applications, 290*, 129–135.
Giuffrida, A., & Boseley, S. (2020). *Italy's pandemic plan old and inadequate, covid report finds. outdated guidelines and lack of protocols may have led to thousands of extra deaths*. The Guardian, August 13, 2020. https://www.theguardian.com/world/2020/aug/13/italy-pandemic-plan-was-old-and-inadequate-covid-report-finds
von Goethe, J. W. (1989). *Goethe's Satyros and Prometheus* (1819). In J. Gray & A. Tille (Eds.). Bauermeister.
von Goethe, J. W. (2012). *The sorrows of young Werther* (1774). In D. Constantine (Ed.). Oxford University Press.
von Goethe, J. W. (2014). *Faust. A tragedy* (1790) (M. Greenberg, Trans.). Yale University Press.
Gogoll, J., & Müller, J. F. (2017). Autonomous cars. In favor of a mandatory ethics setting. *Science and Engineering Ethics, 23*, 681–700.
Goldman, A. I. (2006). *Simulating minds. The philosophy, psychology, and neuroscience of mindreading*. Oxford University Press.
Golumbia, D. (2009). *The cultural logic of computation*. Harvard University Press.
Goodall, N. J. (2014a). Ethical decision making during automated vehicle crashes. *Journal of the Transportation Research Board, 2424*, 58–65.
Goodall, N. J. (2014b). Machine ethics and automated vehicles. In G. Meyer & S. Beiker (Eds.), *Road vehicle automation* (pp. 93–102). Springer.

Greenfield, A. (2018). *Radical technologies. The design of everyday life*. Verso.
Grierson, J., Milmo, D., & Farah, H. (2021). *Revealed: Anti-vaccine TikTok videos being viewed by children as young as nine. Covid misinformation remains on site for months adding to concern over impact of social media on young people*. The Guardian, October 8, 2021. https://www.theguardian.com/technology/2021/oct/08/revealed-anti-vaccine-tiktok-videos-viewed-children-as-young-as-nine-covid
Gurney, J. K. (2016). Crashing into the unknown. An examination of crash-optimization algorithms through the two lanes of ethics and law. *Albany Law Review, 79*, 183–267.
Guyer, P. (2003). Kant on the theory and practice of autonomy. In E. Frankel Paul, F. Miller, & J. Paul (Eds.), *Autonomy*. Cambridge University Press.
Hacking, I. (1990). *The taming of chance*. Cambridge University Press.
Han, B.-C. (2015). *The burnout society*. Stanford University Press.
Hansen, S. B. (2020). Albert Borgmann. Local technologies. In *Philosophers of technology* (pp. 122–148). De Gruyter.
Hansson, S. O. (1996). Decision making under great uncertainty. *Philosophy of the Social Sciences, 3*(26), 369–386.
Hansson, S. O. (Ed.). (2018). *Technology and mathematics. Philosophical and historical investigations*. Springer.
Harari, Y. N. (2018). *21 lessons for the 21st century*. Jonathan Cape.
Harari, Y. N. (2016). *Homo deus. A brief history of tomorrow*. Harvill Secker.
Haworth, L. (1986). *Autonomy. An essay in philosophical psychology and ethics*. Yale University Press.
Heal, J. (2003). *Mind, reason and imagination*. Cambridge University Press.
Hegel, G. W. F. (1975). *Aesthetics. Lectures on fine art*. In T. M. Knox (Ed.). Clarendon Press.
Hegel, G. W. F. (1991). *Elements of the philosophy of right* (H. B. Nisbet, Trans.). In A. W. Wood (Ed.). Cambridge University Press.
Hegel, G. W. F. (2011). *Lectures on the philosophy of world history*. In R. F. Brown & P. C. Hodgson (Eds.)., with the assistance of W. G. Guess. Clarendon Press.
Herzen, A. (1956). *From the other shore* (M. Budberg, Trans.), introduction by I. Berlin. Weidenfeld & Nicolson.
Hesiod. *Theogony* (G. Nagy & J. Banks, Trans.). Centre for Hellenic Studies, Harvard University. https://chs.harvard.edu/primary-source/hesiod-theogony-sb/
Hesiod. *Works and days* (G. Nagy, Trans.). Centre for Hellenic Studies, Harvard University. https://chs.harvard.edu/primary-source/hesiod-works-and-days-sb/
Hevelke, A., & Nida-Rümelin, J. (2015). Responsibility for crashes of autonomous vehicles. An ethical analysis. *Science and Engineering Ethics, 21*, 619–630.
Heyen, N. B. (2020). From self-tracking to self-expertise. The production of self-related knowledge by doing personal science. *Public Understanding of Science, 2*(29), 124–138.
Hight, M. A. (2008). *Idea and ontology. An essay in early modern metaphysics of ideas*. The Pennsylvania State University Press.
Hill, T. (1991). *Autonomy and self-respect*. Cambridge University Press.
Hill, T. (2013). Kantian autonomy and contemporary ideas of autonomy. In O. Sensen (Ed.), *Kant on moral autonomy* (pp. 13–21). Cambridge University Press.
Hill, T. (1989). The Kantian conception of autonomy. In J. Christman (Ed.), *The inner citadel. Essays on individual autonomy* (pp. 91–105). Oxford University Press.
Hill, T. (2000). *Respect, pluralism and justice. Kantian perspectives*. Oxford University Press.
Himmelreich, J. (2018). Never mind the trolley. The ethics of autonomous vehicles in mundane situations. *Ethical Theory and Moral Practice, 21*, 669–684.
Holmes, F. (2020). *Private test tracks are where autonomous vehicles drive risk-free*. Automotive world. https://www.automotiveworld.com/articles/private-test-tracks-are-where-autonomous-vehicles-drive-risk-free/
Horizon 2020 Commission Expert Group to advise on specific ethical issues raised by driverless mobility (E03659) (2020). *Ethics of connected and automated vehicles. Recommendations on*

road safety, privacy, fairness, explainability and responsibility. Publication Office of the European Union.
Horning, B. (2013). *Google alert for the soul*. The new inquiry. https://thenewinquiry.com/google-alert-for-the-soul/
Huxely, A. (1938). *Ends and means. An inquiry into the nature of ideals and into the methods employed for their realization*. Chatto & Windus.
Iacoboni, J. (2018). *L'esperimento. Inchiesta sul Movimento 5 stelle*. Laterza.
Ihde, D. (1979). *Technics and praxis*. Reidel.
Ihde, D. (2001). *Bodies in technology*. University of Minnesota Press.
Ihde, D. (1993). *Postphenomenology. Essays in the postmodern context*. Northwestern University Press.
Ihde, D. (2009). *Postphenomenology and technoscience. The Peking university lectures*. Albany State University of New York Press.
Ilyenkov, E. (2012). Dialectics of the ideal. *Historical Materialism, 2*(20), 149–193.
Jackson-Wright, Q. *Attention engineers. These are the soft skills hiring managers want you to have*. https://www.themuse.com/advice/engineers-soft-skills-hiring-managers-want
Jobin, A., Ienca, M., & Vayena, E. (2019). The global landscape of AI ethics guidelines. *Nature Machine Intelligence, 1*, 389–399.
Jullien, F. (2009). *L'invention de l'ideal et le destin de l'Europe*. Seuil.
Jun, N. J., & Wahl, S. (Eds.). (2010). *New perspectives on anarchism*. Lanham-Boulder-New The Rowman & Littlefield Publishing Group.
Jun, N. J. (Ed.). (2017). *Brill's companion to anarchism and philosophy*. Brill.
Kahn, A. D. (1970). Every art possessed by man comes from Prometheus. The Greek tragedians and science and technology. *Technology and Culture, 2*(11), 133–162.
Kale, S. (2021). *Chakras, crystals and conspiracy theories. How the wellness industry turned its back on covid science*. The Guardian, November 11, 2021. https://www.theguardian.com/world/2021/nov/11/injecting-poison-will-never-make-you-healthy-how-the-wellness-industry-turned-its-back-on-covid-science
Kant, I. (2005). *Critique of pure reason* (1781). In P. Guyer & A. W. Wood (Eds.). Cambridge University Press.
Kant, I. (1998). *Groundwork of the metaphysics of morals* (1785). In M. J. Gregor (Ed.). Cambridge University Press.
Kant, I. (1996). *Critique of practical reason* (1788). In M. J. Gregor (Ed.). Introduction by A. Wood. Cambridge University Press.
Kant, I. (2000). *Critique of the power of judgment* (1790). In P. Guyer (Ed.). Cambridge University Press.
Kekes, J. (2020). *Wisdom*. Oxford University Press.
Kekes, J. (1983). Wisdom. *American Philosophical Quarterly, 3*(20), 277–286.
Kelly, K. (2010). *What technology wants*. Penguin.
Kelly, K. (2012). *The quantified century*. http://quantifiedself.com/conference/Palo-Alto-2012
Keynes, J. M. (1979). The general theory and after. Defence and development. In *The collected writings of John Maynard Keynes*, vol. 14. Macmillan.
Killmister, S. (2017). *Taking the measure of autonomy. A four-dimensional theory of self-governance*. Routledge.
Kind, A., & Kung, P. (Eds.). (2016). *Knowledge through imagination*. Oxford University Press.
Kind, A. (2022). *Imagination and creative thinking*. Cambridge University Press.
Kind, A. (2018). How imagination gives rise to knowledge. In F. Macpherson & F. Dorsch (Eds.), *Perceptual memory and perceptual imagination*. Oxford University Press.
Kitchin, R. (2014). *The data revolution. Big data, open data, data infrastructures and their consequences*. Sage.
Kohler, E. (1953–54). Je ne sais quoi. Ein Kapitel aus der Begriffsgeschichte des Unbegreiflichen. *Romanistisches Jahrbuch, VI*, 21–59.
Korsgaard, C. M. (1996a). *Creating the kingdom of ends*. Cambridge University Press.

Korsgaard, C. M. (1996b). *The sources of normativity*. Cambridge University Press.
Kroes, P. (2012). *Technical artefacts. Creations of mind and matter: A philosophy of engineering design*. Springer.
Kuokkanen, M. (Ed.). (1994). Idealization VII. Structuralism, idealization and approximation. *Poznań Studies in the Philosophy of the Sciences and the Humanities, 42*.
Kurutz, S. (2012). *Skipping the architect. Wise or otherwise?* The New York Times, June 20, 2012. https://www.nytimes.com/2012/06/21/garden/computer-programs-help-users-bypass-the-architect.html
Kurzweil, R. (2005). *The singularity is near. When humans transcend biology*. Viking.
Larmore, C. (2008). *The autonomy of morality*. Cambridge University Press.
Latour, B. (1993). *We have never been modern*. Harvard University Press.
Latour, B. (1999). *Pandora's hope*. Harvard University Press.
Latour, B. (1987). *Science in action. How to follow scientists and engineers through society*. Open University Press.
Leben, D. (2017). A Rawlsian algorithm for autonomous vehicles. *Ethics and Information Technology, 19*, 107–115.
Lehrer, K., et al., (Eds.). (1996). *Knowledge, teaching, and wisdom*. Kluwer Academic Publishers.
Lehrer, K. (1997). *Self-trust. A study in reason, knowledge, and autonomy*. Oxford University Press.
Lehrer, K. (2003). Reason and autonomy. *Social Philosophy and Policy*, *2*(20), 177–198.
von Leibniz, G. W. (1989). Meditations on knowledge, truth, and ideas. In R. Ariew & D. Garber (Eds.), *Philosophical essays* (1684). Hackett.
von Leibniz, G. W. (1981). Preface. In P. Remnant & J. Bennett (Eds.), *New essays on human understanding* (1704). Cambridge University Press.
von Leibniz, G. W. (1988). Felicity. In P. Riley (Ed.), *Leibniz. Political writings* (1694–98). Cambridge University Press.
Leonard, T. (2009). *Hewlett-Packard webcams "racist"*. https://www.telegraph.co.uk/technology/news/6874153/Hewlett-Packard-webcams-racist.html
Lin, P. (2015). Why ethics matters for autonomous cars. In E. Maurer, et al. (Eds.), *Autonomous driving. Technical, legal and social aspects* (pp. 69–85). Springer.
Lindgren, S. (2017). *Digital media and society*. Sage.
Liu, H. Y. (2017). Irresponsibilities, inequalities and injustice for autonomous vehicles. *Ethics and Information Technology, 19*, 193–207.
Lohr, S. (2018). *Facial recognition is accurate, if you're a white guy*. https://www.nytimes.com/2018/02/09/technology/facial-recognition-race-artificial-intelligence.html
London, A. J., & Danks, D. (2018). Regulating autonomous vehicles. A policy proposal. In *Proceedings of the 2018 AAAI/ACM conference on AI, ethics, and society* (pp. 216–221).
Longo, G. O. (2001). *Homo technologicus*. Meltemi.
Lupton, D. (2016a). The diverse domains of quantified selves. Self-tracking modes and dataveillance. *Economy and Society*, *1*(45), 101–122.
Lupton, D. (2016b). *The quantified self*. Polity Press.
Lupton, D. (2012). M-health and health promotion. The digital cyborg and surveillance society. *Social Theory & Health*, *3*(10), 229–244.
Lupton, D. (2018). How do data come to matter? Living and becoming with personal data. *Big Data & Society, 2*, 1–11.
Luttrell, R. (2021). *Social media and society. An introduction to the mass media landscape*. The Rowman & Littlefield Publishing Group.
Lynch, R., & Farrington, C. (Eds.). (2017). *Quantified lives and vital data. Health, technology and society*. Palgrave Macmillan.
Marcus Aurelius (1800). Meditations of Marcus Aurelius Antoninus, Emperor of the Romans. G. Long (Ed.). Lee and Shepard.
Marshall, P. H. (2010). *Demanding the impossible. A history of anarchism*. PM Press.
Martin, B., & Brouwer, W. (1993). Exploring personal science. *Science Education*, *4*(77), 441–459.

Martin, G. A. (2020). Sustainability interrogation of the autonomous vehicle at its society-technology interface. *Bulletin of Science, Technology & Society, 3–4*(39), 23–32.

Matthias, A. (2004). The responsibility gap. Ascribing responsibility for the actions of learning automata. *Ethics and Information Technology, 6*, 175–183 (2004)

Maxwell, N. (1984). *From knowledge to wisdom. A revolution for science and the humanities*. Basil Blackwell.

May, T. (1994a). The concept of autonomy. *American Philosophical Quarterly, 2*(31), 133–144.

May, T. (1994b). *The political philosophy of poststructuralist anarchism*. Pennsylvania State University Press.

Mayor, A. (2018). *Gods and robots. Myths, machines and ancient dreams of technology*. Princeton University Press.

McDermott, D. (2017). Why ethics is a high hurdle for AI (2008). In W. Wallach & P. Asaro (Eds.), *Machine ethics and robot ethics*. Routledge, chapter 20.

McLaughlin, P. (2007). *Anarchism and authority. A philosophical introduction to classical anarchism*. Routledge.

Mecacci, G., & Santoni de Sio, F. (2020). Meaningful human control as reason-responsiveness. The case of dual-mode vehicles. *Ethics and Information Technology, 22*, 103–115.

Mele, A. (1991). History and personal autonomy. *Canadian Journal of Philosophy, 23*, 271–280.

Mele, A. (1995). *Autonomous agents. From self-control to autonomy*. Oxford University Press.

Meyers, D. T. (1989). *Self, society, and personal choice*. Columbia University Press.

Mill, J. S. (1956). *On liberty (1859)*. The Liberal Arts Press.

Millar, J. (2015). Technology as moral proxy. Autonomy and paternalism by design. *IEEE Technology and Society Magazine, 34*, 47–55.

Millar, J. (2016). An ethics evaluation tool for automating ethical decision-making in robots and self-driving cars. *Applied Artificial Intelligence, 8*(30), 787–809.

Miller, T. (2019). Explanation in artificial intelligence. Insights from the social sciences. *Artificial Intelligence, 267*, 1–38.

Mirza, R. D., Punja, S., Vohra, S., & Guyatt, G. (2017). The history and development of N-of-1 trials. *Journal of the Royal Society of Medicine, 8*(110), 330–340.

Mittelstadt, B. D., Allo, P., Taddeo, M., Wachter, S., & Floridi, L. (2016). The ethics of algorithms. Mapping the debate. *Big Data & Society, 2*(3), 1–21.

Moor, J. (2006). The nature, importance and difficulty of machine ethics. *IEEE Intelligent Systems, 4*(21), 18–21.

Morozov, E. (2013). *To save everything click here*. Allen Lane.

Morton, T. (Ed.). (2002). *A Routledge literary sourcebook on Mary Shelley's Frankenstein*. Routledge.

Murray, J. (1833). *A classical manual, being a mythological, historical, and geographical commentary on Pope's homer and Dryden's Virgil*. Murray.

Nafus, D., & Sherman, J. (2014). This one does not go up to 11. The quantified self movement as an alternative big data practice. *International Journal of Communication, 8*, 1784–1794.

Nafus, D. (2016). *Quantified. Biosensing technologies in everyday life*. MIT Press.

Nagel, T. (1986). The problem of autonomy. In *The view from nowhere* (pp. 110–120). Oxford University Press.

Nagy, P., Wylie, R., Eschrich, J., & Finn, E. (2020). Facing the Pariah of science. The Frankenstein myth as a social and ethical reference for scientists. *Science and Engineering Ethics, 2*(26), 737–759.

Neff, G., & Nafus, D. (2016). *Self-tracking*. MIT Press.

Newman, S. (2016). *Postanarchism*. Polity Press.

Nichols, S., & Stich, S. P. (2003). *Mindreading. An integrated account of pretense, self-awareness and understanding other minds*. Oxford University Press.

Nissenbaum, H. (1996). Accountability in a computerized society. *Science and Engineering Ethics, 1*(2), 25–42.

Noble, S. U. (2018). *Algorithms of oppression. How search engines reinforce racism*. New York University Press.

Nowakowa, I., & Nowak, L. (Eds.). (2000). Idealization X. The richness of idealization. *Poznań Studies in the Philosophy of the Sciences and the Humanities, 69*.

Nozick, R. (1974). *Anarchy, state, utopia*. Basic Books.

Nozick, R. (1989). *What is wisdom and why do philosophers love it so? In the examined life* (pp. 267–278). Touchstone Press.

Nunes, A., Reimer, B., & Coughlin, J. F. (2018). People must retain control of autonomous vehicles. *Nature, 556*, 169–171.

Nyholm, S. (2018a). The ethics of crashes with self-driving cars. A roadmap (I and II). *Philosophy Compass, 13*, 1–10.

Nyholm, S. (2018b). Attributing agency to automated systems. reflections on human-robot collaborations and responsibility-loci. *Science and Engineering Ethics, 24*, 1201–1219.

O'Neil, C. (2016). *Weapons of math destruction. How big data increases inequality and threatens democracy*. Crown.

Official Journal of the European Union, 58 (2015), C 421/2, https://eur-lex.europa.eu/legal-content/EN/TXT/?uri=OJ%3AC%3A2015%3A421%3AFULL

Oshana, M. (1998). Personal autonomy and society. *Journal of Social Philosophy, 1*(29), 81–102.

Ovid. (2010). *Metamorphoses* (S. Lombardo, Trans.), introduction by W. R. Johnson, Hackett Publishing Company.

Owens, J., & Cribb, A. (2019). My fitbit thinks I can do better! Do health promoting wearable technologies support personal autonomy? *Philosophy & Technology, 32*, 23–38.

Palmirani, M., & Sapienza, S. (2021). Big Data, explanations and knowability. *Ragion pratica, 2*, 349–364.

Panofsky, D., & Panofsky, E. (1978). *Pandora's box. The changing aspects of a mythical symbol*. Princeton University Press.

Paruchuri, V. *Happsee*. https://quantifiedself.com/show-and-tell/?project=723

Pasquale, F. (2015). *The black box society*. Harvard University Press.

Pirandello, L. (1992). One, no one and one hundred thousand (1925). W. Weaver (Ed.). Sprul Editions.

Plato. (1921). Theaetetus. In *Plato in twelve volumes*, vol. 12 (H. N. Fowler, Trans.). Harvard University Press.

Plato. (1925). Symposium. In *Plato in twelve volumes*, vol. 9 (H. N. Fowler, Trans.). Harvard University Press.

Plato. (1956). *Protagoras* (B. Jowett, Trans.), revised by M. Ostwald, G. Vlastos (Eds.). Bobbs-Merrill.

Plato. (1966). Apology. In *Plato in twelve volumes*, vol. 1 (H. N. Fowler, Trans.). Harvard University Press.

Plato. (1967). Protagoras. In *Plato in twelve volumes*, vol. 3 (W. R. M. Lamb, Trans.). Harvard University Press.

Plato (1969). *Respublica*. In P. Shorey (Ed.). Harvard University Press.

Plato (2009). *Phaedrus*. In R. Waterfield (Ed.). Oxford University Press.

Polanyi, M. (1958). *Personal knowledge. Towards a post-critical philosophy*. University of Chicago Press.

Porter, T. M. (1995). *Trust in numbers. The pursuit of objectivity in science and public life*. Princeton University Press.

Powers, T. (2006). Prospects for a Kantian machine. *IEEE Intelligent Systems, 21*, 46–51.

Powers, T. (2013). Prospects for a Smithian machine. In *Proceedings of the International Association for Computing and Philosophy*, College Park.

Purves, D., et al. (2015). Autonomous machines, moral judgment, and acting for the right reasons. *Ethical Theory and Moral Practice, 18*, 851–872.

Quarteroni, A. (2021). *Algoritmi per un nuovo mondo*. Edizioni Dedalo.

Rawls, J. (1999). *A theory of justice. Revised edition* (1971). Harvard University Press.

Raz, J. (1986). *The morality of freedom*. Oxford University Press.
Reath, A. (2006). *Agency and autonomy in Kant's moral theory. Selected essays*. Oxford University Press.
Rees, M. (2018). *On the future. Prospects for humanity*. Princeton University Press.
Reichenbach, H. (1959). *The rise of scientific philosophy (1951)*. University of California Press.
Rescher, N. (1987). *Ethical idealism. An inquiry into the nature and function of ideals*. University of California Press.
Rifkin, J. (1995). *The end of work. The decline of the global labour force and the dawn of the post-market era*. Putnam Publishing Group.
Roberts, S. *QS + Paleo = ?* https://quantifiedself.com/show-and-tell/?project=206
Roff, H. M., & Moyes, R. (2016). *Meaningful human control, artificial intelligence and autonomous weapons*. Briefing paper prepared for the Informal Meeting of Experts on Lethal Autonomous Weapons Systems, UN Convention on Certain Conventional Weapons, article 36, pp. 1–6.
Rorty, R. (1982). *Consequences of pragmatism. Essays, 1972–80*. University of Minnesota Press.
Rorty, R. (1989). *Contingency, irony, and solidarity*. Cambridge University Press.
Rosa, H. (2010). *Alienation and acceleration. Towards a critical theory of late-modern temporality*. NSU Press.
Rose, A. (2010). *Are face-detection cameras racist?* http://content.time.com/time/business/article/0,8599,1954643,00.html
Rosenberg, N. (1994). *Exploring the black box. Technology, economics and history*. Cambridge University Press.
Rosenberger, R., & Verbeek, P. P. (Eds.). (2015). *Postphenomenological investigations. Essays on human-technology relations*. Lexington Books-The Rowman & Littlefield Publishing Group.
Rosenfeld, A., & Richardson, A. (2019). Explainability in human-agent systems. *Autonomous Agents and Multi-Agent Systems, 33*, 673–705.
Ross, D. (1951). *Plato's theory of ideas*. Clarendon.
Ross, A. (2016). *The industries of the future*. Simon & Schuster.
Rousseau, J.-J. (1978). *Emile. Or on education* (1762). In A. Bloom (Ed.). Basic Books.
Ruckenstein, M., & Pantzar, M. (2017a). Beyond the quantified self. Thematic exploration of a dataistic paradigm. *New Media & Society, 19*(3), 401–418.
Ryan, S. (1999). What is wisdom? *Philosophical Studies, 93*, 119–139.
Ryan, S. (2012). Wisdom, knowledge and rationality. *Acta Analytica, 2*(27), 99–112.
Santoni de Sio, F. (2017). Killing by autonomous vehicles and the legal doctrine of necessity. *Ethical Theory and Moral Practice, 20*, 411–429.
Santoni de Sio, F., & van den Hoven, J. (2018). Meaningful human control over autonomous systems. A philosophical account. *Frontiers in Robotics and AI, 5*, 1–14.
Santoni de Sio, F. (2021). The European commission report on ethics of connected and automated vehicles and the future of ethics of transportation. *Ethics and Information Technology, 23*, 713–726.
Santoro, G. (2012). *Un Grillo qualunque. Il Movimento a 5 stelle e il populismo digitale nella crisi dei partiti italiani*. Castelvecchi.
Sartwell, C. (2008). *Against the state. An introduction to anarchist political theory*. SUNY Press.
Schafer, K. (2018). A brief history of rationality. Reason, reasonableness, rationality, and reasons. *Manuscrito, 4*(41), 501–529.
Schenk, H. G. (1966). *The mind of the European romantics*. Constable.
Schiller, J. C. F. (1844). The ideal and the actual life. In *The poems and ballads of Schiller* (Sir. E. Bulwer Lytton, Trans.). Tauchnitz.
Schiller, J. C. F. (1910). Letters upon the aesthetic education of man. In *Literary and philosophical essays. French, German and Italian*. Collier, The Harvard Classics.
Schneewind, J. B. (1998). *The invention of autonomy*. Cambridge University Press.
Schüll, N. D. (2018). Self in the loop. Bits, patterns, and pathways in the quantified self. In Z. Papacharissi (Ed.), *A networked self and human augmentics, artificial intelligence, sentience* (pp. 25–38). Taylor & Francis.

Selke, S. (Ed.). (2016). *Lifelogging. Digital self-tracking and lifelogging. Between disruptive technology and cultural transformation*. Springer.
Sen, A. (1999). *Development as freedom*. Oxford University Press.
Seneca. (1928). To Serenus on the firmness of the wise man. In J. W. Basore (Ed.), *Moral essays*. William Heinemann LTD.
Seneca. (2007). In B. Inwood (Ed.), *Selected philosophical letters*. Oxford University Press.
Sensen, O. (Ed.). (2013). *Kant on moral autonomy*. Cambridge University Press.
Shakespeare, W. (1998). *Hamlet*. Oxford University Press.
Shariff, A., Bonnefon, J. F., & Rahwan, I. (2017). Psychological roadblocks to the adoption of self-driving vehicles. *Nature Human Behaviour*, *1*(10), 694–696.
Sharon, T. (2017) Self-tracking for health and the quantified self. Re-articulating autonomy, solidarity, and authenticity in an age of personalized healthcare. *Philosophy & Technology*, *30*, 93–121.
Sharon, T., & Zandbergen, D. (2017). From data fetishism to quantifying selves. Self-tracking practices and the other values of data. *New Media & Society*, *11*(19), 1695–1709.
Shellabarger, N. *28 years of tracking, but what have i learned?* https://quantifiedself.com/show-and-tell/?project=815
Shelley, P. B. (1968). *Prometheus unbound. The text and the drafts: Toward a modern definitive edition*. In L. J. Zillman (Ed.). Yale University Press.
Shelley, M. (1994). *Frankenstein. Or, the modern Prometheus* (1818 and 1831). Penguin Books.
Shrader-Frechette, K., & Westra, L. (Eds.). (1997). *Technology and values*. The Rowman & Littlefield Publishing Group.
Singh Chawla, D. (2019). *Elsevier investigates hundreds of peer reviewers for manipulating citations. The publisher is scrutinizing researchers who might be inappropriately using the review process to promote their own work*. https://www.nature.com/articles/d41586-019-02639-9
Smarr, L. *Where there's data there's hope*. https://quantifiedself.com/show-and-tell/?project=390
Smith, W. (Ed.). (1867). *Dictionary of Greek and Roman biography and mythology*. Little, Brown & Co.
Smith, D. (2018). *Exceptional technologies. A continental philosophy of technology*. Bloomsbury.
Sophocles. (1981). *Œdipus the king (OT)* (F. Storr, Trans.). Harvard University Press.
Stayton, E., & Stilgoe, J. (2020). It's time to rethink levels of automation for self-driving vehicles. *IEEE Technology and Society Magazine*, 13–19.
Steiner, G. (2001). *Grammars of creation. Originating in the Gifford lectures for 1990*. New Yale University Press.
Stendhal. (1949). *Memoirs of an egotist* (1832) (T. W. Earp, Trans.). Turnstile Press.
Stokes, D. R. (2016). Imagination and creativity. In A. Kind (Ed.), *The Routledge handbook of philosophy of imagination* (pp. 247–261). Routledge.
Stueber, K. (2011). Imagination, empathy, and moral deliberation. The case of imaginative resistance. *The Southern Journal of Philosophy*, *49*, 156–180.
Sullins, J. P. (2016). Artificial phronesis and the social robot. *Frontiers in Artificial Intelligence and Applications*, *290*, 37–39.
Swan, M. (2013). The quantified self. Fundamental disruption in big data science and biological discovery. *Big Data*, *1*(2), 85–98.
Sysling, F. (2019). Measurement, self-tracking and the history of science. An introduction. *History of Science*, *2*(58), 103–116.
Taiebat, M., et al. (2018). A review on energy, environmental, and sustainability implications of connected and automated vehicles. *Environmental Science & Technology*, *20*(52), 11449–11465.
Tatarkiewicz, W. (1980). *A history of six ideas. An essay in aesthetics* (1975) (C. Kasparek, Trans.). The Hague-Boston, Nijhoff-Kluwer.
Taylor, C. (1991). *The ethics of authenticity*. Harvard University Press.
Taylor, J. S. (Ed.). (2005). *Personal autonomy. New essays on personal autonomy and its role in contemporary moral philosophy*. Cambridge University Press.
Taylor, R. (2005). Kantian personal autonomy. *Political Theory*, *5*(33), 602–628.

Tegmark, M. (2017). *Life 3.0. Being human in the age of artificial intelligence.* Allen Lane.
Tiberius, V. (2008). *The reflective life. Living wisely with our limits.* Oxford University Press.
Till, C. (2019). Creating "automatic subjects". Corporate wellness and self-tracking. *Health, 4*(23), 418–435.
Timmer, J. *My life in 40 variables.* https://quantifiedself.com/show-and-tell/?project=847
Tobias, B. (2019). *Understand, manage, and prevent algorithmic bias. A guide for business users and data scientists.* Apress.
Torres, P. (2017). *Morality, foresight, and human flourishing. An introduction to existential risks.* Pitchstone Publishing.
Tronconi, F. (Ed.). (2015). *Beppe Grillo's Five Star Movement. Organisation, communication and ideology.* Farnham, Ashgate.
Turing, A. M. (1950). Computing machinery and intelligence. *Mind, 49*, 433–460.
Turner, F. (2006). *From counterculture to cyberculture. Stewart brand, the whole earth network, and the rise of digital utopianism.* University of Chicago Press.
United States Department of Defense (2012). Autonomy in weapon systems. *Directive 3000*(09), 1–15.
Vaidhyanathan, S. (2011). *The googlization of everything (and why we should worry).* University of California Press.
Vallor, S. (2015). Moral deskilling and upskilling in a new machine age. Reflections on the ambiguous future of character. *Philosophy & Technology, 28*, 107–124.
Vallor, S. (2016). *Technology and the virtues. A philosophical guide to a future worth wanting.* Oxford University Press.
Vallor, S. (2017). AI and the automation of wisdom. In T. M. Powers (Ed.), *Philosophy and computing. Essays in epistemology, philosophy of mind, logic, and ethics* (pp. 161–178). Springer.
van de Poel, I., & Goldberg, D. E. (Eds.). (2010). *Philosophy and engineering. An emerging agenda.* Springer.
van de Poel, I., & Royakkers, L. (2011). *Ethics, technology, and engineering. An introduction.* Wiley-Blackwell.
van de Poel, I., Royakkers, L., & Zwart, S. D. (2015). *Ethics, moral responsibility and the problem of many hand.* Routledge.
Verbeek, P. P. (2005). *What things do. Philosophical reflections on technology, agency, and design.* Pennsylvania State University Press.
Verbeek, P. P. (2011). *Moralizing technology. Understanding and designing the morality of things.* University of Chicago Press.
Vernant, J.-P. (1980). *Myth and society in ancient Greece* (J. Brighton, Trans.). Harvester Press.
Walker Rettberg, J. (2014). *Seeing ourselves through technology. How we use selfies, blogs and wearable devices to see and shape ourselves.* Palgrave Macmillan.
Wallach, W., & Allen, C. (2008). *Moral machines. Teaching robots right from wrong.* Oxford University Press.
Wallach, W., & Asaro, P. (Eds.). (2017). *Machine ethics and robot ethics.* Routledge.
Walton, K. L. (1990). *Mimesis as make-believe.* Harvard University Press.
Watson, G. (1975). Free action. *The Journal of Philosophy, 72*, 205–220.
Waymo Safety Report (2017). *On the road to fully self-driving.* https://www.auto-mat.ch/wAssets/docs/171019_waymo-safety-report-2017-10.pdf
Waymo Safety Report (2021). *On the road to fully self-driving.* https://storage.googleapis.com/waymo-uploads/files/documents/safety/2021-08-waymo-safety-report.pdf
Weizenbaum, J. (1976). *Computer power and human reason. From judgment to calculation.* Freeman.
Whitcomb, D. (2010). Wisdom. In S. Bernecker & D. Pritchard (Eds.), *Routledge companion to epistemology.* Routledge.
Williams, K. (2013). *The weight of things lost. Self-knowledge and personal informatics* (pp. 1–4). CHI.

Winner, L. (1977). *Autonomous technology. Technics-out-of-control as a theme in political thought.* MIT Press.

Wittgenstein, L. (1969). *On certainty.* Blackwell.

Wittgenstein, L. (1963). *Tractatus logico-philosophicus* (1921). In D. F. Pears & B. F. McGuinness (Eds.). Routledge & Kegan Paul.

Wolf, G. I., & De Groot, M. (2020). A conceptual framework for personal science. *Frontier in Computer Science*, 2(21), 1–5.

Wolf, G. I. (2009). *Know thyself. Tracking every facet of life, from sleep to mood to pain, 24/7/365.* Wired. http://archive.wired.com/medtech/health/magazine/17-07/lbnp_knowthyself?currentPage=all

Wolf, G. I. (2010). *The data-driven life.* The New York Times Magazine. http://www.nytimes.com/2010/05/02/magazine/02self-measurement-t.html?pagewanted=all&_r=0

Wolff, R. P. (1970). *In defence of anarchism.* Harper and Row.

Xu, W. (2021). From automation to autonomy and autonomous vehicles. Challenges and opportunities for human-computer interaction. *Interactions, 1*, 49–53.

Zagzebski, L. (1996). *Virtues of the mind. An inquiry into the nature of virtue and the ethical foundations of knowledge.* Cambridge University Press.

Zellini, P. (2018). *La dittatura del calcolo.* Adelphi.

The manufacturer's authorised representative in the EU is Springer
Nature Customer Service Centre GmbH, Europaplatz 3, 69115 Heidelberg,
Germany. If you have any concerns regarding our products, please
contact ProductSafety@springernature.com

Printed and bound by CPI Group (UK) Ltd, Croydon, CR0 4YY
01/10/2026
02225876-0008